THE LEARNING PARADOX

How our unconscious mind
(mis)guide our learning strategies

FELIPE FREGNI

Felipe Fregni, M.D., Ph.D., M.M.Sc., M.P.H., M.Ed.
Professor of Physical Medicine and Rehabilitation, Harvard Medical School
Professor of Epidemiology, Harvard T.H. Chan School of Public Health
Director, Spaulding Neuromodulation Center, Spaulding Rehabilitation Hospital

& Guest author for bonus chapter on research:
Anna Marduy
Spaulding Neuromodulation Center, Spaulding Rehabilitation Hospital, Harvard Medical School

Acknowledgments

I would like to thank my father for the great discussions we had for each chapter of this book. His engineering and university faculty member background was beneficial to clarify some important points of this book to the readers. I would also like to thank the editors who helped edit each chapter of this book for clarity. Finally, I am also grateful to all my students and teaching assistants from the Teaching, Learning, and Innovation (TLI) program for being a great source of inspiration for this book.

TABLE OF CONTENTS

Introduction

"My wife ought not even to be under suspicion" (Julius Caesar, 100 – 44 BC)

During 44-46 B.C., the Roman emperor Julius Caesar was asked why he divorced his wife, Pompeia. He famously said that his wife had to not only be honest but also appear to be honest. Caesar's statement came as the result of an episode involving another political figure, Publius Clodius Pulcher. Pulcher was arrested and tried for trying to seduce Caesar's wife.

Julius Caesar had political reasons for these actions, and this book will not go into whether this was a good political move. However, this book takes the opposite stance to Julius Caesar. One must learn deeply regardless of whether one appears to be learning, such as having excellent grades. It is certainly possible to do both; however, it is crucial to understand the forces leading to deep learning and those leading to the appearance of learning. Although these forces may coincide, they often go in opposite directions.

One of the main issues discussed in this book is that unconscious forces that may drive us away from deep learning are part of the normal cognitive structure guiding our behavior. An important message delivered by this book is that the cognitive forces that guide our behavior and occasionally lead us to make poor decisions are the same ones that create poor learning behavior. This topic has been discussed more frequently in other fields, particularly economics, such as by Dan Ariely and Daniel Kahneman.[1,2]

Dan Ariely is an economics professor and author who has written many books on this subject, such as *Predictably Irrational*, where he discussed interesting examples of students' procrastination.[1] He compared three methods for requesting papers to be graded as part of the main course assignments. One was a traditional method in which the professor determined three deadlines (one for each paper) distributed along the course, which offered no flexibility to students. The second method offered full flexibility, where students could hand in the paper at any time during the course, including handing in all three papers together on the final day. Finally, with the intermediate method, students could choose staggered deadlines, with corresponding penalties for being late, or handing in all three papers at the end of the course. The study found that the method with no flexibility was the best to help students overcome procrastination. Chapter 3 discusses the neural forces at play when deciding between immediate and delayed rewards. In the above example, students were likely to put off writing papers in favor of other activities and leave paper-writing to the last minute, which could result in lateness or not finishing all three papers. In addition, this book discusses cognitive mechanisms to decrease the forces causing procrastination. Although it may seem counterintuitive, the above study indicated that deadlines were necessary to counteract procrastination (but there are also other methods to deal with procrastination as discussed in Chapter 7).

Another famous example of a study on hidden cognitive forces that guide behavior was conducted by the Nobel Prize winner Daniel Kahneman.[2] He presented several examples of the cognitive system demonstrating the irrationality of some decisions. He created a simple human cognition model, dividing the cognitive structure into two main systems: fast and slow thinking. Fast thinking represents subcortical processing with little conscious awareness, which is responsible for most thinking. This system is fast, intuitive, driven by emotions, and usually associated with errors. Slow thinking is critical and analytical thinking, which requires more time to process. Some cognitive errors that impact learning may be due to the fast and automatic thinking system making inefficient decisions during the learning process, which is discussed throughout this book.

The following chapters address how irrational decisions can negatively affect learning. These are referred to here as irrational decisions, as we may perceive learning as efficient when making these decisions.

I would also like to mention a personal debate, which I often discussed with my father—an active university professor throughout his entire life. He often mentioned that humans learn naturally, which is a central point of this book. Humans learn naturally and have a sophisticated neural system to deeply learn complex topics. However, a system is only as good as the operator that manages it. The fact that humans learn naturally does not prevent us from misusing (or misguiding) our learning machinery.

My father would argue that humans do not need a manual to use their brains to learn. If a two-year-old can learn something as complex as a human language without a manual, humans should be able to learn naturally. This book argues that people created systems, such as traditional schools, to learn, which are undoubtedly useful, particularly considering public policy. However, these systems may hinder the natural ability to learn deeply. Furthermore, even with these systems in place, there are many truly outstanding individuals who stand out despite the force of the education system. This book questions whether it is possible to have more outstanding individuals or for these individuals to be even more outstanding. This book argues that there is no ceiling to humans' ability to learn deeply.

However, admittedly, one point was difficult to resolve during the writing of this book—that is, the book's audience. The readers of this book are likely to be aware of this topic and seeking to gain a deeper understanding. Readers are likely already trying to understand the learning process as a critical step in becoming a better learner. A book such as this may not reach those who are unaware of learning processes and their importance. However, some readers may be a transformation agent. Educators may pass along the notion that this and other books discuss the understanding of learning mechanisms.

An important consideration is that this book does not provide an exhaustive analysis of the hidden cognitive forces that may hinder our natural ability to learn deeply. It is my hope that this book will inspire readers to consider and learn more regarding these topics. Finally, these hidden cognitive biases may not have the same effect on everyone, and this effect may not have a meaningful educational effect for everyone. Educational research to date has not addressed the predictors of these effects and effect sizes. The last chapter of this book discusses existing research that aimed to understand these effects through experiments in order to provide inspiration for further reading.

Felipe Fregni

Boston, Massachusetts

References for introduction

1. Ariely, D. (2010). *Predictably Irrational, Revised and Expanded Edition: The Hidden Forces That Shape Our Decisions*(Revised and Expanded ed.). Harper Perennial.
2. Kahneman, D. (2011). *Thinking, Fast and Slow*. Farrar, Straus and Giroux.

Chapter 2: The Ivy League Paradox

Students cram day and night during high school, work on their extra-curricular activities, prepare their applications for an Ivy Leave school, and finally, perhaps, get accepted. They continue their hard work to achieve an almost perfect grade point average (GPA), a common metric used in the US and some other countries to measure average grade. They study as much as they can to do well in exams and course activities. When graduation comes, some will be among the top performers in their group. On graduation day, they may think: "My work is done. I can relax and enjoy life. I will soon be hired by a big firm and start a successful career."

Not so soon. An almost perfect GPA from an Ivy League school will certainly help a graduate to get a number of interviews; however, their grades as measured by the GPA metric do not predict whether they will be hired. Interestingly, these findings are not new. Two scientists published a detailed analysis in 1983 to determine the factors during the hiring process that would lead to increased productivity. They showed that the only important factor was the cognitive ability of the candidates assessed for the job. Other factors, including grading during school, have very low relevance in determining a candidate's fitness to a position.[1,2]

This is the paradox of learners relying on grading to predict future job or life success. A perfect GPA certainly shows great qualities, such as discipline, to follow all the courses, dedication, and responsibility. These parameters are undoubtedly important. But they do not represent the ability to learn in different environments quickly and to attain different skills in an unstructured manner.

From medical school to research laboratory: the feeling of being lost

Feeling lost is a problem that I observe very often in my laboratory at Harvard—much more often than I would like. Usually, several students apply to spend a year or more in my laboratory learning clinical research as the first steps toward becoming a clinical scientist. Most of them have recently graduated as physicians from different countries. They usually have only a formal education, going from high school to medical school, and then to my laboratory. They often come from prestigious schools in their home countries and are excellent students there.

An interesting common phenomenon for most of them is a sense of disorientation when they arrive. Of course, they arrive from different countries, and the initial adaption is indeed difficult; however, the sense of freedom they have in my laboratory makes them a bit disoriented. They are used to being told what to do, what to study, when to study, and to being given the exact blueprint to obtain "good grades." However, becoming a scientist is the opposite of formal education. Scientists need to think independently, to study independently and autonomously, and to be creative to resolve the questions they have formulated. In addition, learning has changed in the digital world. The individual has become connected to a node in a large network of knowledge facilitated by digital methods. Therefore, the learner depends on finding the information necessary to improve his understanding. This is the theory of connectivism, as proposed by Siemens.

This is a conflict for my young, recently arrived fellows. They expect to be told what time to arrive, what lectures they need to attend, what to do in our clinical research studies, what they need to read, and how they are tested. This does not happen. In the laboratory, they are assigned to be part of clinical studies and they must learn in the process. We discuss these studies in our meetings and encourage them to go deep and study all aspects of each study. However, this is difficult for them to achieve. Their learning habits have not developed. They know very well how to study for an exam, but not to think independently and creatively.

One interesting aspect is that most of them are hard workers and want to be assigned to many laboratory activities. So, the main point that I insist on in the laboratory is to give them time to learn about each study.

This is one of the consequences of how the traditional school system worsens students' learning efficiency. When they are confronted with a real situation, they fail to learn independently and be creative. This is not a fixed condition, obviously. Over time, students learn how to think critically, and some become great scientists. However, this takes time and is much more complicated than it should be.

(Un)learning statistics in graduate courses

Imagine the following situation. You decide to move to a new house, which has a different layout to your previous house. There are no specific spaces, such as a kitchen, bedroom, or bathroom. The house is an open space with a bunch of file cabinets with several drawers and spaces to keep all your stuff. You arrive with all your boxes, and within 3 days, everything is organized beautifully. However, you have not organized all of your objects with a special rationale. You simply kept opening the boxes and storing your items in the first open space you saw.

To make matters worse, you go traveling and return only after a few years. That night, you want to find a wine bottle opener. It seems like an impossible mission. Likely, you will need to open all the drawers randomly to find it.

How does this story relate to students learning statistics? At college, students are focused on memorization: for example, a parametric statistical test is used to compare two groups that measure the outcome using continuous and normally distributed variables. They learn exactly what is needed for the exam and do not delve deeper into the topic, to try to understand the concepts further, or to understand how what is learned would impact their future careers. Therefore, learning goes to one of the student's random "brain drawers", and this memory will likely fade away with time, as there is no key to access this learning. The learning was designed only to pass the exam.

The result of such learning at college is that, if this student becomes a physician in the future and then reads a scientific paper whose authors measure pain levels on a continuous scale versus a categorical scale, the student will not remember what that means for understanding the results, the statistics of the scientific paper, and so on.

It could be argued that such learning involves the students training their working memory, which is an advantageous cognitive function to be developed for future professional life. Although it is not disadvantageous to have excellent working memory performance, it brings little additional advantage. This is especially true in our current society, in which access to information is almost effortless. In this context, it is important to process and analyze information instead of memorizing it. It would be unusual for a boss to ask an employee to go home and memorize three chapters of a recent book on leadership and then to respond to a written exam the next day.

Zipf's law: the law of least effort

George Kingsley Zipf was an American linguist born in 1902. He was one of the first linguists to study word frequency. He observed that by ranking the frequency of the words, one could find an exponential decay for the total time the words are used according to their rank. In other words, the second most frequent word is used half of the time compared to the most frequent word. The third most frequent word is used one-third of the time, and so on. This phenomenon seems to be applicable to other examples, such as corporation size, income, and even people watching the same TV channel.

There is still no acceptable mechanism to explain Zipf's law. One potential explanation is the principle of least effort. Based on this principle, to make ourselves understood, we would use the words that require the least effort—in this case, the most frequent words.

The principle of least effort is intuitive. A biological system seeks to save energy and complete a task with the least amount of both time and effort. This seems to be the default mode; indeed, it is highly adaptive.

For learning, our biological behavior does not seem to be different. We also tend to conserve energy. In the same way, students aim to maximize their grading scores with the least amount of energy and time. If they can obtain an A+ without the need to pay attention in class or even missing classes completely and studying only the day before the exam, there is no reason that most students would not do so.

It is possible that top students are doing both: aiming to learn deeply in their courses and, at the same time, aiming for A+. Even then, the deficiencies in our education system are evident, as most courses are not designed to give a student autonomy to go beyond what is taught or to question what is being taught and the ideas of the instructor. Doing so would certainly result in a lower grade. It is then likely that this behavior of seeking top grades leads to less overall analytical thinking and creativity skills.

Survivor: brains vs. brawn vs. beauty

Survivor is a successful TV show that is now in its 38th season. The show's format is based on about 20 participants being left on an isolated island or in another location; for 38 days, they need to outlast their colleagues to win a USD 1 million prize. The elimination of participants is decided by their peers (tribe members). The game involves a significant amount of cognitive and social thinking. The challenge in the game is for participants to balance helping their colleagues while appearing to be weak to avoid being eliminated while making alliances to remain in the game. Ultimately, it is a real-life game that involves constant strategic thinking.

In one of the seasons, the producers decided to choose participants based on three main characteristics: (a) "Brains": participants in this group were selected based on jobs that require high academic achievement; (b) "Brawn": participants in this group were selected based on jobs that demand high physical competitiveness; and (c) Beauty":

participants were chosen based on their beauty and charisma.

In this season, participants from the Brawn group were the most successful, two out of three reached the three finalist stages, and the winner was from this tribe. The survivor, a police officer from Boston, said in an interview that he was thinking about strategy day and night.

This TV show is, of course, a complex scenario, as the game allows deception, and it is difficult to translate such complexity to real life. However, this reality show has an interesting feature that applies to our discussion: the survivor game is a new situation for participants and depends on their high learning skills. They need to learn in a short amount of time the characteristics of other participants and how to adapt to new situations quickly. The interesting part of this show is to observe precisely how dynamic the contest is.

For example, high past achievement in a specific academic area that is likely artificial and does not represent real life does not reflect future success in the job market. This is especially true nowadays with rapidly changing markets. Companies are interested in hiring recruits who can learn different domains in a short amount of time, or who can add value to our society.

Goldman Sachs metric of hiring: neither school nor high GPA

The financial services company Goldman Sachs is one of the top investment banks and one of the largest organizations in the U.S. A position at Goldman Sachs is a dream job for many graduates. Although the company's hiring algorithm is not fully disclosed publicly, interviews with human resources representatives and analyses of the company's hires show that an employee's university or GPA and gradings during school years have very little influence on being hired. Although a very low GPA and academic performance might not secure a participant an interview, a high GPA has little weight on the candidate's possible hiring.

In an interview with CNBC, Dane Holmes, Goldman Sachs' head of human capital management, listed some essential characteristics that the company's recruiters are looking for when interviewing candidates.[3]

1. A potential employee might not even have a major in finance: there is no perfect educational background for this company. While this seems counterintuitive, it in fact has exactly the opposite outcome: if candidates have great ability to learn, they will quickly learn the needed financial background.
2. Candidates need grit, to show that they follow their passion and are passionate about pursuing their goals. Passion significantly enhances learning.
3. Candidates should recognize and learn from failure. Although learning from mistakes should be considered a given, as the way we learn is through trial and error, the cognitive insights of learning from failures need to be integral to the learning of the individual.
4. Participants who let demonstrate emotion when telling stories during an interview are

considered honest and able to give important weight to emotion in their professional life, which is important for learning.

5. Interviewees should show a balance between work and life/hobbies. An important characteristic of deep learning is to work intensively on a project for some time and then to disconnect from it. This allows other brain circuits to be activated. This is one of the best strategies for developing creative solutions to a problem.

These characteristics make it clear that Goldman Sachs values new learning. The company values learning that will have a significant impact on the institution rather than learning that makes a teacher happy.

Do you want to work at Google? Stop paying attention to your grades

Google is another dream firm for many recent and not-so-recent graduates. It is a leading technology company and one of the largest global companies, with constant innovation in the technology field.

Google's senior vice president of people operations, Laszlo Bock, stated in an interview with the *New York Times*: "GPAs are worthless as criteria for hiring, and test scores are worthless." In this interview, Laszlo admitted that unless a candidate was fresh out of college, the company might not even ask for the GPA.[4]

What is essential for Google when hiring new recruits? Similar to Goldman Sachs, Google is interested in real-life learning rather than learning achieved by artificial exam scores. Google has called its method a behavioral interview, meaning that it is interested in learning how a potential employee reacts and adapts to real-life situations. The company even hires employees without a college education if the candidate has other required qualities.

As a company focused on innovation, Google seeks to hire people who can drive innovation and resolve complex problems in a world constantly changing at a rapid pace.

The vicious and virtuous cycles: why do ineffective learners become more ineffective and effective learners more effective?

Poverty is a classic example of a vicious cycle. The poor have little or no money to study and learn. They have to start working at an early age. In addition, poor individuals usually borrow money at high interest rates and even regularly pay late fee penalties, thereby becoming poorer. Rich individuals are on the opposite spectrum. They can invest more time and money in additional learning. They can also invest their money to generate interest and thus, over time, they make more money and become richer.

The same dual cycle applies to learners focused on grading versus learners focused on real and meaningful learning. An individual that is conditioned to focus on grades learns how to be effective in memorizing and doing well in exams. As expected, learning becomes a tedious task. Over time, the person becomes less interested in learning and is more interested in getting satisfactory grades only and proceeding to the next step in the education system. Learning becomes a means to achieve the student's goals and not an end in itself. Thus, over time, the individual's deep and effective learning worsens.

The opposite is also true: an individual focused on deep and critical learning, over time, becomes more effective, connects information quickly, and learns more effectively. As the individual learns more, there is a growth in learning-related brain neural circuitry, which improves learning efficiency.

Low fat–high carbohydrate diet to prevent cardiovascular disease and high grades for a good education

The last topic in this chapter addresses an essential issue in our society: public policy recommendations. Usually, public policy is based on strong scientific evidence, right? Not always. When scientific development is in its initial stage, there is a reasonable risk that the evidence is not sufficiently reliable. The initial studies are either biased (meaning their results are not valid) or represent random variations that produce contrary results.

Take the example of cardiologists, the American Heart Association, and the U.S. government recommending a diet low in fat and high in carbohydrates. In the 1960s, scientists believed that the plaque in the arteries leading to a heart attack could be caused by excessive fat consumption. The research was in its initial stages at the time. Posterior evidence has shown that the recommendation is wrong, and in fact, on the contrary, could lead to a heart attack. By contrast, a diet high in sugars may lead to low-grade inflammation in the gut, leading to an increased immune system reaction that also damages the artery walls.

One important issue is that there may be only a limited amount of evidence needed to generate a new public policy at the beginning, especially if there is no other policy. However, it is much harder to change an existing policy, as habits have formed and there is more evidence required.

We have been brainwashed since day 1 of our schooling that grading is what matters. Parents review their children's grades and give them "carrots and sticks" according to their results. This starts in elementary school, and continues through middle school, high school, and graduate school, when the emphasis on grading starts to change; however, at this point, our belief in A+ as the perfect surrogate for learning is entrenched.

In my book, *Critical Thinking in Teaching and Learning,* I discuss methods for assessing students that do not rely (or that rely less) on grading to measure knowledge; however, the shift in attitude needs to come from students.

References for chapter 2

1. Hunter, J. E., & Schmidt, F. L. (1983). Quantifying the effects of psychological interventions on

employee job performance and work-force productivity. *American Psychologist*, *38*(4), 473–478. https://doi.org/10.1037/0003-066x.38.4.473

2. Russiello, M. (2016, January 16). Does college GPA predict job performance? www.linkedin.com. https://www.linkedin.com/pulse/does-college-gpa-predict-job-performance-mike-russiello/
3. Connley, C. (2019, April 3). How to get a job at Goldman Sachs. CNBC.Com. https://www.cnbc.com/2019/04/02/suzy-welch-how-to-get-a-job-at-goldman-sachs.html
4. Umoh, R. (2018, July 31). This is the only time you should include a GPA on your resume. CNBC.Com. https://www.cnbc.com/2018/07/31/when-to-include-your-gpa-on-your-resume.html

Chapter 3 - Are There Two Marshmallows for the Learner? The Irresistible Immediate Gratification

You may have heard about the famous experiment testing delayed gratification in children. The results of this experiment were first published by psychologists Walter Mischel and Ebbe B. Ebbesen at Stanford University in 1970[1]. In this experiment, they offered two options to children of ages 3 to 5 years:

Option 1: You can now have a reward immediately (for instance, a cookie or a marshmallow). Researchers used different rewards such as cookies, pretzels, and marshmallows.
Option 2: If you can wait 15 minutes, you can have two rewards (for instance, another cookie or another marshmallow).

In the experiment, the following instruction was given to these children:

"Let's see what's under here. I'll bet it's a surprise. Oh boy, look at that A marshmallow and a pretzel. Which would

you like to eat? You can eat either the marshmallow or the pretzel [At this point, the child chose which one he wanted to eat]. Oh, you know what? I have to go out of the room now, and if you wait until I come back by myself then you can eat this one [pointing to chosen object] right up. But, you know, if you don't want to wait you can ring the bell and bring me back anytime you want to. But if you ring the bell then you can't have this one, but you can have that one (pointing to the unchosen object) So, if you ring the bell and bring me back then you can't have the ___________, but you can have
the _____________."

How did the children do? As you can imagine, it was hard to wait for 15 minutes to get more gratification. Interestingly, in this experiment, the researchers observed that the children's behavior during the experiment contributed to the outcome (e.g., waiting or not for the delayed gratification). Children who were trying to distract themselves, by playing, singing, or moving in their chair seem to be able to resist, for a longer time, the temptation of getting one marshmallow immediately rather than wait till the end to get two marshmallows.

In fact, in a follow-up experiment, the same group of researchers compared different experimental conditions that modulated their attention off the reward. In one of them, they instructed the child to play with a toy during the waiting time. In the other one, they instructed the child to think about a fun experience during the waiting time. The results were very interesting. In the control group (no distractor), children waited less than 1/2 minute on an average, and none of them could wait for the 15-minute waiting time. In contrast, the average waiting time for the toy and cognitive distractor was much higher: 8.59 and 12.12 minutes, respectively. Moreover, four children in the toy group and six in the cognitive distractor group could wait for the full 15 minutes.[2]

Now, back to the classroom. What are the incentives students are offered to study the course material deeply? None or very little. This is a fundamental problem for the learner. There is very little incentive for critical and deep thinking.

What happens in a traditional class? The instructor walks in the classroom and provides a syllabus with a long list of topics to be studied and hundreds of pages to be read. The instructor does say that if you read all the material and do well in the exam and other course activities, you will get "one marshmallow," that is, the A+ grade. If you learn what the instructor wants you to and put the minimum necessary effort only to study the syllabus, you do get your full marshmallow.

But do you get a second marshmallow if you go beyond what the teacher is asking and learn deeply about that topic? No. At least not directly from the course. Additionally, we are exposed to a massive number of distractors (i.e., other activities, hobbies, etc.). There are hundreds of different types of marshmallows that are not educational and can be obtained immediately. In this chapter, I will discuss these two factors: the lack of incentives for long-lasting and deep learning, and the competing incentives to detract the learner from deep learning.

Why do students cheat? The fastest way to get the marshmallow

In my global training clinical research program, which is now offered by the Harvard T.H. Chan School of Public Health, there are students from several countries.

Students gather locally in a classroom to connect via video-conference with our Boston site, for instance, from Madrid, Spain, Bogota, Colombia, or Sao Paulo, Brazil. We call the students in the same local group as a course site (i.e., students from Dresden, Germany are in the Dresden site). This program is divided into modules, and after every module, students undergo an exam. Although this exam composes very little of their final grade—which they are aware of— they feel pressured to get a good grade in this Harvard course. To decrease the pressure, I allow students to bring their books and notes, making it an open book exam (we currently entirely changed the method of assessing the students, but this past experience provides interesting insights). Initially these exams were conducted on-site with all students from the same site taking the exam together. Most of my students are physicians, recently graduated from medical school, and in some sites, there are senior physicians. Do you think, in this case, students would cheat? How about if it is a site with more senior physicians?

Although, I cannot make generalizations as this was not a formal experiment, my experience was that several students did cheat on the exams. How did I find this out? It was not difficult. Our site monitor may have been too friendly to the students in two sites and did not truly monitor the exam well, since results on these two sites were odd. All students in these two sites (about 15 students) had the same multiple-choice answers in an exam of about 25 questions. Though this is not impossible, it is extremely unlikely that this would occur by chance. Moreover, in our other sites where we knew the site monitor was taking his role more seriously, there was a large variation. Furthermore, the exam is moderately difficult (making it unlikely to be the ceiling effect of an easy exam).

Interestingly, this also happened in one of our sites in which students were all relatively senior specialized physicians with an average of 10 or more years of clinical practice experience after residency. We had to change the format and method of the exam altogether to avoid this. But why does this happen?

The answer is multifactorial, but there is one important contributor: the incentive. What incentive is the student is looking for? The A+ grade. Cheating in exams is one of the strongest examples of how students behave in this incentive system.

Going to our example of marshmallows, imagine the following situation. Students are given a multiple-choice test, and if they do well, they get their marshmallows. However, there is a twist—students correct the exams themselves, and you allow them to shred their answer sheet. They only tell you how many of their answers were correct answers. What would they do in this situation? Dan Ariely tried to answer this question through an experiment.

Harvard MBA students who cheated in the exam

In the book *Predictably Irrational*, cognitive psychologist Dan Ariely writes about his elegant experiment with Harvard MBA students.[3] As you may have guessed , it was no coincidence to select MBA students from the prestigious Harvard Business School. This is a top school in its field, and the alumni lead a significant number of 500 fortune companies.

The experiment was simple and straightforward. In the control group, students had to answer 50 general factual knowledge questions (for instance, "what is the longest river in the world"). They then gave the exam to the co-investigator in the room, and the right answers were counted. Participants got 10 cents for each right question. Participants, on an average, gave 32.6 correct answers in the exam. In this group, there was no (at least easy) opportunity to cheat.

The interesting part comes now. Ariely and his colleagues created three experimental groups to tempt the students. In the first of these groups, participants, at the end of the exam, graded their exam using a grading sheet. They returned the exam and the grading sheet to the co-investigator and collected their money. Here, therefore, participants could mark a question they had wrong as a correct one.

In the other two groups, the temptation to cheat was even higher. In the second group, participants destroyed the evidence (they shredded their exams) and then returned the grading to the proctor and collected their money. In the third group, they did not even need to face the examiner; they destroyed both the exam and grading sheet and simply collected the money from a big jar. It is important to note that the students in all three groups were given the rules in the beginning of the experiment.

What happened? The results were clear. When given the chance to collect more marshmallows (money in this case), the MBA students took the chance and collected more money. These three experimental groups had a higher number of correct answers (on an average, 36.2, 35.9, and 36.1 in these three groups, respectively, or about four questions more than the control group). Interestingly, results were similar across the three groups.

What do these results tell us? Participants cheat to get more immediate rewards, especially when the reward is directly correlated to the final grade. It is important to remember that high achieving students likely have great academic skills and did not have to cheat to get where they were. This seems similar to the senior physicians that did not need to cheat to pass my program. But they still chose to do it.

However, there is a significant difference here. One can argue that there was no incentive for participants to behave ethically as they would not learn more if they studied more on this exam. However, in a regular program, what if we offer the second marshmallow if they study more to get a better grade without cheating?

The secret to getting the second marshmallow is to get distracted from the first marshmallow

In the Stanford experiment, the only way children could wait longer for the delayed gratification (i.e., the second candy) was to either distract themselves with a toy or engage themselves in another positive thought. When this was done, the time children waited increased significantly, and approximately 50% of them could wait the full 15 minutes to get the second candy.

One insight from this study on human incentives is where the attention is focused. If students are focused only on grading, their attentional resources will be directed to the exam and to get that reward maximized, similar to the children when there was no other cognitive distractor. In this case, the average time was less than 1/2 minute, indicating that they waited for almost no time before they claimed their one but only reward.

Do educational programs take the focus away from the exams? Does the school system take away the focus from tests? Do parents take away the focus from grading score sheets of their children? No, schools, parents, and society always send a clear message: grading is what is essential. The consequence is simple: students will concentrate on that reward (i.e., getting an A+). Thus, their strategy is to maximize their gains in the most time efficient manner. But, another question is whether students truly believe there is a second marshmallow out there for them.

However, one consideration here is that students may be intrinsically motivated to study mathematics, for instance. In this case, the marshmallow (or reward) comes from the sense of mastery. These students would keep studying mathematics to get the reward of resolving difficult and challenging math questions. However, our current system insists on having grading and exams as the main marshmallow.

The expectation of a second marshmallow affects how hard you work for it

In a follow-up experiment by the University of Rochester, a group of scientists tested another factor that can influence the decision to work for delayed gratification (i.e., waiting to get the second marshmallow). In this experiment, the scientists made one important modification. Before the experiment, children were exposed to a similar situation where they could learn whether they could trust the researchers.[4]

For half of the children, the researchers showed to be unreliable as they came back to their room and said, in this training condition, "I am sorry, but unfortunately I could not find the "second marshmallow" (in this case it was a set of crayons). For the other half, they fulfilled their promise and brought the extra reward as promised.

These two groups of children were then exposed to a similar marshmallow experiment. The results were unequivocal. The group that did not believe that the second marshmallow would come had a mean waiting time of 3 min and 2 s, while in the group that learned they could trust the researcher, the waiting time was significantly higher (average of 12 min and 2 s).[4]

These results demonstrate that not only self-control is important to work for a reward but also how much you believe that you will receive this reward if you do work hard. What does this mean for learning beyond getting an A+ in a multiple-choice exam? Students need to truly believe that there is a large reward coming if they go the extra mile to learn the topic more deeply.

The challenge in real-life education is even more significant. In the experiment, the children knew the reward was coming quickly. In real-life, if students are more dedicated and try to learn deeply about a topic, the second marshmallow is their success in life, which is a very delayed reward. Moreover, it is not guaranteed that success will come, such as in the unreliable group in the marshmallow experiment where children decided not to wait for the second reward.

CME credits for physician's learning: marshmallows are not only for children

The scenario is no better for adult learning. While I agree that adults do have more self-control, their brain architecture is not very different from that of a 5-year-old child. Let us go back to our example of cheating in exams. My experience with the physicians in my course shows as well as Ariely's study with MBA Harvard students confirms that participants cheat for immediate rewards.

Adult learning is indeed very challenging. One example is the medical profession. Physicians need to obtain several continuing medical education (or CME) credits to renew their licenses. For instance, in Massachusetts, physicians must get at least 50 CME credits to renew their licenses. Medical societies do not rely on physicians to learn by themselves.

This is a great initiative, right? Physicians need to be as updated as possible to provide the best clinical care. The principle is not wrong; however, again, the incentive is problematic. CME credits are given for participation in a conference or for completing a course. Medical conferences usually have very low educational efficiency. These events cover several topics in very short and non-interactive lectures. Participants get tired very easily with this marathon, and indeed, attendance is generally low, especially toward the end of the conference. It is similar to reading dozens of book synopses in a short amount of time and calling it learning. Most participants attend only for the sake of the credits. Here, I would not even blame the learner as it would be challenging to engage in meaningful learning in these events.

If medical accreditation societies do not trust their members—physicians—who have spent so many years in medical school and residency, to learn by themselves, the problem is likely worse in other professions. Moreover, the methods (such as CME credits to physicians) to address this lack of self-learning are still providing wrong incentives to the learners.

The second marshmallow is an elusive one

Let us go back to the marshmallow experiment. Remember when the Stanford team decided to compare the self-control of preschool children by introducing distractors? The results were unequivocal. If there were no cognitive distractors, children focused only on getting the first reward and getting it done.

Similarly, in education, if there is no cognitive awareness of learning, students would only focus on the first marshmallow. The goal becomes to get the A+ as efficiently as possible.

Now, let us imagine the following modification to the experiment. I will adapt the instructions of the Stanford experiment (my new text is in bold—the rest is the original text from the experiment):

"Let's see what's under here. I'll bet it's a surprise. Oh boy, look at **this delicious marshmallow**. Oh, you know what? I have to go out of the room now, and if you wait until I come back by myself then you can eat this marshmallow and**, in addition, you will learn that working harder to get a marshmallow will bring several future benefits to you as you will learn to be patient and your parents will be proud of**

you, etc., etc. But, you know, if you don't want to wait you can ring the bell and bring me back anytime you want to. But if you ring the bell you will not be able to learn this important lesson, but you can have that marshmallow."

What is different here? The investigator made the child aware of the benefit of the learning. The second marshmallow is the learning itself. The key here is to provide enough intrinsic value for that marshmallow to our learners. Sounds easy, right? Not so much though, since our attention system tends to focus on what may provide more benefits in the short term.

Which would you prefer: 1,000 dollars now or 1,000 dollars in 3 years?

The answer is obvious for most of us. Most of us would prefer money now. If someone offers you 1,000 dollars in 3 years, this likely means this person is now offering less than 1,000 dollars. There are a few reasons money is more valuable now than in the future. To start with, you can invest 1,000 dollars now, and in three years, this will incur interests, and you will likely get about 100 dollars or more in interest depending on your investments. Money now means this money can be used to buy things now and receiving the reward immediately rather than in three years from now.

The same pattern of behavior occurs for the learner. It is truly difficult to study for very delayed gratification. To make matters worse, this gratification may not even come. There is no guarantee that if you spend the evening reading a full book in your subject, making notes, and truly reflecting on it, this will lead to additional success in your job. It is likely that it will, but there may be too much uncertainty in this bet for many.

That is why the system of providing the first marshmallow (the grades, the diploma, and so on) for students that study for the exam and get a satisfactory grade works relatively well to keep students attending colleges, going to courses, and so forth. The reward is immediate. You get your 1,000 dollars now. The question that is rarely discussed is how much this "1,000-dollar" reward is valued.

It is difficult for the human brain to focus on something else than the first marshmallow. This is especially true when it requires spending more time and energy for a second marshmallow that may never come and may be so late that its actual value would only be a small piece of marshmallow if given now. For the learner, the rational decision is to take the first marshmallow now and then go after another marshmallow (i.e., another course).

The endowment effect and Steven Jobs

The Nobel prize winner Daniel Kahneman and his colleagues discussed the endowment effect with a series of examples in their review article.[5] The endowment effect means that our value for giving up an object is much higher than the value of the amount to acquire the same object. This is related to our loss aversion. Losing something we have is more costly for us than buying the same object. Kahneman and his colleague give the example of a bottle of wine whose price increased significantly with time. For instance, a wine lover purchased a bottle of wine for 10 dollars, and over the years, the wine price increased dramatically; it is now worth 200 dollars. However, this person would neither sell this wine for 200 dollars, nor buy it for that price.

Another example is a study in which participants received either 2 dollars or a lottery ticket. When participants with the lottery ticket were offered 2 dollars to sell their tickets, very few accepted. Similarly, in the group with the ones who received 2 dollars, very few agreed to trade the money for the lottery ticket. Although this would not make sense, this is related to our loss aversion behavior.

Now imagine you are halfway through college, and you realized that this education is not what you are looking for or that your learning has been very poor. Would you give up then? Giving up would mean that you would lose everything you paid up to that point and also lose all the credits you acquired. Most would not give up but a few would. Steven Jobs was one of them. In his 2005 Stanford Commencement speech[6], he said:

"After six months, I couldn't see the value in it. I had no idea what i wanted to do with my life and no idea how college is going to help me figure it out. So i decided to drop out and trust that it would all work out OK. It was

pretty scary at the time, but looking back it was one of the best decisions I ever made. I decided to take Calligraphy class to learn how to do this. I learned about Serif and Sans Serif typefaces, about varying the amount of space between different letter combinations, about what makes great typography great.
None of this had even a hope of any practical application in my life.
But after 10 years later, when we were designing the first Macintosh computer, it all came back to me.
It was the first computer with beautiful typography.
If i had never dropped in that single course in college, the Mac would have never had multiple typefaces or proportionally spaced fonts.
And since Windows just copied the Mac, it's likely that no personal computer would have them."

As you can see above, Jobs was getting rewards from his intrinsic motivation system. In fact, when learning becomes about getting a grade and a diploma, this creates the opposite effect of truly excelling in someone's skills and passions. This behavior of chasing an A+ may even worsen the cycle of poor learning (discussed in chapter 2). Even those who are aware that the learning is not effective may continue the course or program to avoid losing the effort and funds invested and becoming an even worse learner. Would it not be wonderful if our educational system emphasizes the intrinsic motivation system instead of only offering the immediate reward from extrinsic motivators (i.e., the "marshmallows")?

References for chapter 3

1. Mischel, W., & Ebbesen, E. B. (1970). Attention in delay of gratification. *Journal of Personality and Social Psychology*, *16*(2), 329–337. https://doi.org/10.1037/h0029815
2. Mischel, W., Ebbesen, E. B., & Raskoff Zeiss, A. (1972). Cognitive and attentional mechanisms in delay of gratification. *Journal of Personality and Social Psychology*, *21*(2), 204–218. https://doi.org/10.1037/h0032198
3. Ariely, D. (2010). *Predictably Irrational, Revised and Expanded Edition: The Hidden Forces That*

Shape Our Decisions(Revised and Expanded ed.). Harper Perennial.

4. Kidd, C., Palmeri, H., & Aslin, R. N. (2013). Rational snacking: Young children's decision-making on the marshmallow task is moderated by beliefs about environmental reliability. *Cognition*, *126*(1), 109–114. https://doi.org/10.1016/j.cognition.2012.08.004
5. Kahneman, D., Knetsch, J. L., & Thaler, R. H. (1991). Anomalies: The Endowment Effect, Loss Aversion, and Status Quo Bias. *Journal of Economic Perspectives*, *5*(1), 193–206. https://doi.org/10.1257/jep.5.1.193
6. Stanford University. (2017, June 12). *Text of Steve Jobs' Commencement address (2005)*. Stanford News. https://news.stanford.edu/2005/06/14/jobs-061505/

Chapter 4 – Why Choosing a Coffee-maker Makes Us Learn Less: Effects of Cognitive Dissonance on Learning

You are asked to rate the desirability of eight objects: all new-looking household items in the same price range. They are, for instance, an automatic coffee-maker, an electric sandwich grill, a silk-screen reproduction, an automatic toaster, a fluorescent desk lamp, a book of art reproductions, a stopwatch, and a portable radio. Your rating must evaluate their attractiveness and usefulness. In an interesting twist, you then receive two items from this list at random and are asked to choose just one of them. At this point, you are probably wondering whether this is some sort of trick, or if you will simply be given the one you rated higher and take your prize home. Thankfully, the latter is true. However, the experiment offers more than an unexpected reward. Your choice here has implications that reveal a great deal about your thinking process. This information, in turn, has implications for education.

The aforementioned scenario was taken from Jack Brehm's study at the University of Minnesota. Published in 1956[1], it was one of the first studies on cognitive dissonance. In this study, the first part of the experiment involved 225 female students (this is the stage in which participants rated each of the 10 objects). After that, two randomly chosen items were presented, and participants chose one to take home. Here it must be noted that the items *appeared* to be random selections from the perspective of the participants. In truth, the options were not randomly compiled, but randomized to be high, medium, and low dissonance.

High dissonance means that the desirability of the choices given to participants was close (for instance, the second- and third-most desirable objects). On the other hand, low dissonance occurs when the two choices are far apart on the desirability scale (for instance, the second- and the eighth-most desirable objects). The underlying logic here is that when a participant chooses one object, they automatically lose the other – a scenario that produces an internal conflict or a dissonance. Conversely, when one object is deemed significantly superior to its competitor, there is neither conflict nor dissonance.

So how do people manage scenarios when dissonance is high (meaning one must abandon an option they really liked – perhaps only slightly less than the other)? This is simple: one modifies their original reasoning regarding the two objects as a means to reduce this dissonance, a phenomenon that was evident in the Brehm study. The results were unequivocal. When participants were asked to rate the objects again, the high dissonance group decreased the ratings of the unchosen options significantly compared to the chosen ones. It was the only group to do so. In the low dissonance condition, the change in the second rating was negligible.

Here, it must be noted that participants did not know the true goals of this experiment, and believed they were helping the manufacturer rate its products. Moreover, they were told that the second desirability rating served to help the manufacturers understand what transpired once someone had purchased one of the products. Therefore, this cognitive adjustment clearly occurred as an internal mechanism that was beyond the individual's awareness. It means that the choice was not explicitly or deliberately made.

Let us suppose a participant were given a toaster and a coffee-maker and she chose the coffee-maker (this is a high dissonance condition). This choice results in a deep-seated need to modify her thoughts about the one not chosen (i.e., the toaster) to decrease its desirability. This makes her feel better about the chosen one (i.e., the coffee-maker).

When students join an educational program, they often encounter two-choice decisions. For instance, the option to study their class materials or go out with friends, or engage in extensive, planned, and consistent study throughout the course versus cramming the day before the final exam. The previous chapters discuss various factors that do not encourage the choice of in-depth study. In applying the previous discussion to this case, it is evident that in-depth study would likely receive a consistently lower rating than the alternatives. Accordingly, this chapter discusses what happens when that scenario creates cognitive dissonance for the learner. Indeed, this is a significant consideration that is likely to be overlooked.

A student who cares more about learning is more affected by cognitive dissonance

Returning to Brehm's article, the participants in the high dissonance group (the second option was also very desirable) not only decreased the desirability of the second option in the second rating but also increased the desirability of the chosen object. This resulted in an enhanced and significant gap between the ratings of the two items.

Consider the following scenario. A responsible and capable student who cares about academics and wants to excel. The student is exposed to other options than in-depth study of the course material. These options may be playing soccer with friends as a member of a school team, traveling, or any other competing engagement or pursuit. At some point, circumstances may shift in favor of these engagements over in-depth study. Moreover, there is the attractive option of cramming at the last minute to get a good grade.

According to the theory of cognitive dissonance, this responsible student will experience high dissonance. Choosing soccer instead of studying math will produce an internal conflict. To resolve this, he employs a simple cognitive trick. He undergoes a reasoning process that increases the desirability of soccer and decreases the desirability of studying math. For instance, he begins to rationalize that the game is an exercise activity, making it important for his health. Furthermore, he continuously emphasizes the point that he can cram before the exam and still get a good grade. In this manner, he reconfigures his thoughts to convince himself that he can swiftly memorize the course material and get a good grade. Moreover, a good grade – even without in-depth learning – would provide him external validation that he is a good student. He then changes his cognitive stance to accommodate his decision, thereby reinforcing the belief that in-depth study throughout the course is less desirable. In this manner, he is freed from the conflict because he feels better for choosing a more desirable option.

Choosing between a 2020 BMW and a 1995 Honda Civic will not affect the desirability of either car

Unless you are a fan of antique cars or you do not like the BMW, choosing a new luxury car should not generate conflict or high dissonance. In this case, you would simply choose the brand-new BMW, get your car, and forget about the 25-year-old Honda. Moreover, you would not like the Honda Civic any less because of this decision. This is, according to the findings of Brehm's study, what determined that the participants offered two objects with massive disparity in desirability ratings did not change their assessment of either object after making the choice.

Of course, this revelation can be logically extended to students who are not as invested in academics as the learner in the previous scenario. In the current case (i.e., students not invested in academics), the desirability scores of both in-depth study and the soccer game will go unchanged once the decision to engage in the latter is made (i.e., these students will not decrease even further the desirability of studying). On the surface, this appears to be a good thing, yet the matter is more complex than that. It is positive in the sense that it would not further reduce the student's desire to engage in deeper study. However, it is not ideal for students to start off with a perception that learning is less desirable. In summary, cognitive dissonance may bring the students who desire deep learning closer to students who do not care about their study habits.

Why do we learn to love courses that result in little meaningful learning?

You may be familiar with the following scenario. It is the first day of a new class you are taking. During the introduction, the teacher explains that there will be many long lectures, hundreds of pages to read, and a protracted final exam. The course begins and, as you feared, the lectures are boring and monotonous. As you have always done, you prepare for the final exam by reading most of the material at the last minute. Consequently, you experience extreme stress in the assessment itself.

Subsequently, someone asks about your experience in the program. It elicits one of two responses.

Option 1: This course was a complete waste of my time. I did not experience meaningful learning as I had to memorize a huge collection of materials for the exam. No one else should take this course.

Option 2: This course is demanding, but it has a robust body of materials. I learned a lot in this program.

While option 1 may immediately appear to be the most likely response, option 2 should not be dismissed. To demonstrate why, it is necessary to consider the findings of Elliot Aronson (University of Stanford) and Judson Mills' (U.S. Army Leadership Human Research Unit) experiment. Published in 1959, the study involved 63 college women.[2] They were told that the experiment served to investigate the "dynamics of group discussion" on the predetermined topic of sex. Moreover, the participants were informed that their physical placement during the session would be pre-set for them.

In the main part of the experiment, the researchers created three groups of stress. One segment of the study population was allocated to the high-stress group, the other to mild stress, and a control group to no stress. The stress exposure involved participants in reading potentially embarrassing material about sex. In the high-stress circumstance, the material was very embarrassing, and in the mild-stress situation, it was much less so.

Participants were told that this part of the experiment (stress exposure) served to introduce them to the group discussion process. After exposure to the varying levels of duress (high, mild, or no stress), participants listened to a tape of the supposed group discussion they were meant to participate in. The taped discussion was specifically designed to be boring and banal. Thus, according to the researchers, *"participants spoke dryly and haltingly on secondary sex behavior in the lower animals, 'inadvertently' contradicted themselves and one another, mumbled several non-sequiturs, started sentences that they never finished, hemmed, hawed, and in general conducted one of the most worthless and uninteresting discussions imaginable."*

At the end of the taped discussion, they were asked to rate the conversation they heard. Significantly, participants rated the bland discussion differently according to the stress they were exposed to. Those exposed to high stress before hearing the tape gave the conversation significantly higher scores than the other two groups (low stress and control/no stress). There was no difference in the ratings of the mild and no stress groups: both were lower than the high-stress group.

Participants in the high-stress group had a high dissonance. They were exposed initially to an uncomfortable situation and then had to listen to a very boring discussion. From this perspective, they had basically suffered the former for nothing. To decrease this conflict, the easiest cognitive response was to increase the discussion's rating to justify the ordeal of their involvement in the initial discussion.

What do you think of the course example from the beginning of this section (the class that was boring and stressful with an exam based on hundreds of pages)? How do you think students may rate that learning experience? One method to decrease the dissonance of going through this stressful experience, only to receive little or no learning, is to rate the course highly.

One important consideration is that this post-event adjustment due to high dissonance will not turn something extremely unsatisfying into the opposite. It will decrease to some extent. However, that may be enough to decrease the high negative experience with courses that were less effective.

Do poor expectations lead to poor grades?

You have most likely heard someone say, "I am not good at math. No matter how much I study, I always fail math tests." This common expression raises the question of whether poor expectation can lead to poor learning outcomes. As this is not a book on positive thinking, this section is more concerned with whether there is a scientific explanation for this relationship. Although the scientific evidence in this area is inconclusive (meaning there are contradicting results), some experiments have posed this exact question. Let us explore the results together.

Elliot Aronson and J. Merrill Carlsmith from Harvard University conducted an interesting experiment to test whether performance expectancy predicts the performance itself.[3] To this end, the researchers first ensured that participants learned about their performance (as determined by the scientists). Next, they created a dissonance condition, tested the subjects, and observed their responses.

Forty female undergraduates were selected to participate. They were told that they were designing an experiment to correlate personality with a type of quiz. In other words, the researchers supposedly wanted to create a quick test that would uncover personality details.

On this premise, the researchers administered this test (a memory card test) four times to each person. Half of the participants were told that their performance was high (about 80% of their answers were correct). The other half was told that their performance was low (about 20% of their answers were right). Then came another manipulation that was perhaps the most important detail of the experiment. Participants received the following results for the final test:

- Ten subjects with high scores received results consistent with their expectations: high scores for the high-high group
- The other ten subjects in the high scores group received results contrary to their expectations: low scores for the high-low group
- Ten subjects with low scores received results consistent with their expectations: low scores for the low-low group
- The other ten subjects of the low scores group received results contrary to their expectation: high scores in the high-low group

The final step of this experiment served to assess whether the groups whose results contradicted their initial expectations would ultimately have different results, compared to those whose results matched their expectations.

The researchers then pretended there was a problem with the final test (inconsistent results) and participants had to repeat them. They could thus test whether the subjects tried to change their responses to match their expectations. What happened then?

Let us first analyze the groups with matching results (low-low and high-high). As expected, these groups had relatively little change in their responses (an average of 6.7 and 3.9 response changes, respectively). However, the groups with dissonant results showed significantly more changes (approximately 11.1 response changes for the high-low group and 10.2 response changes for the low-high group).

Let us analyze the high dissonance group carefully. In this case (the low-high group), one would assume that the participants who scored high in the most recent test while expecting low scores would maintain similar responses. The logic here is that doing so would help them get similarly high scores once again. According to the results, however, this was not the case. These participants changed 10.2 of their responses, on average, which was much more than the average of 3.9 in the high-high group. It means that they likely felt uncomfortable with that dissonance as they expected to have low scores. Similar outcomes were observed in the high-low group.

One alternative explanation for this phenomenon is that the subjects may have wanted to please the researchers by maintaining the same scores. Notably, however, the article mentions that the researchers disclosed the experiment's objective after the fact, and the subjects in the low-high condition were surprised. In their opinion, they did not make their changes with the intention to perform worse. This supports the explanation that the subjects unknowingly tried to get closer to their expectations, even if that meant doing worse in the test.

Does this mean that someone who thinks they are weak at math will only get poor results, given their poor expectations? The answer is not so simple, as there are several contributing factors. Still, the potential of cognitive mechanisms to reduce dissonance between expected and actual results may play a role. This may not be the case concerning all weak students, but it will be for some, perhaps to varying degrees.

How can cognitive dissonance be reduced in education?

The first step is to make it explicit (an additional discussion will be provided in Chapter 6). Students must be made aware of how cognitive dissonance affects their learning. For instance, in the first example, they need to know that when they choose to play soccer (this creates dissonance for the committed students), in particular, the brain will try to resolve this conflict. It will then decrease this dissonance by making the option of studying for an exam less attractive.

Thus, the student should be explicitly aware that choosing to play soccer indeed hurts their learning. A quick note here: my point is not that students should study all the time, but that they need to understand the cognitive conflict taking place. They should then find methods to resolve it without downplaying the value of study time.

In addition, if a student forces himself to study, he can deliberately use cognitive dissonance to make the alternative (playing soccer) less attractive during study time. This technique may also work for some.

How about courses involving high stress and poor learning outcomes? Of course, these should theoretically be modified. However, this is not a realistic expectation across the board. Thus, students should be aware of this, and not classify these experiences as good for their learning. This insight will help the pupil learn more independently, or choose other pathways, like famed entrepreneur Steve Jobs did during his college days. Notably, I am not advocating for students to drop out, but merely stating that learners must critically evaluate their own learning experiences.

Perhaps, most importantly, the learner should know how their brain reacts to learning situations. Many people believe they are in complete control of their mind. However, this is not true. Various cognitive processes unconsciously influence our decision-making process. Yet, because we do have the ability to make decisions, we think we are in control. In actuality, the decisions we arrive at are influenced by criteria that are not fully "disclosed" to us. Thus, the learner's job is to understand their decision-making process and use this knowledge to make better decisions that result in more effective learning experiences.

The final point of this chapter addresses the influence of expectation on learning outcomes. Although I wanted to introduce this topic in this chapter on cognitive dissonance, the topic is equally important and deserves a full chapter to itself. Indeed, expectation does have implications for whether we learn effectively and deeply and will be discussed in the next chapter.

References for chapter 4

1. Brehm, J. W. (1956). Postdecision changes in the desirability of alternatives. *The Journal of Abnormal and Social Psychology*, *52*(3), 384–389. https://doi.org/10.1037/h0041006
2. Aronson, E., & Mills, J. (1959). The effect of severity of initiation on liking for a group. *The Journal of Abnormal and Social Psychology*, *59*(2), 177–181. https://doi.org/10.1037/h0047195
3. Aronson, E., & Carlsmith, J. M. (1962). Performance expectancy as a determinant of actual performance. *The Journal of Abnormal and Social Psychology*, *65*(3), 178–182. https://doi.org/10.1037/h0042291

Chapter 5 – A Sugar Pill for Learning: Is There a Placebo Effect in Learning?

Heart disease has always been one of the major causes of mortality and morbidity. Clinical researchers have always attempted to develop novel treatments to treat this condition. In the 1950s, practitioners began to conduct a less invasive surgery to ligate (i.e., interrupt flow) the mammary artery in the thorax, which was superficial and thus relatively easy to operate upon. The goal was to stop its flow in order to increase blood flow in the coronary arteries (i.e., the arteries that supply blood to the heart). Although some of the experiments at the time did not exhibit this mechanism clearly, this surgery did seem to alleviate heart pain (or angina) significantly. It also had significant benefits: it was a relatively minor surgery.

However, the lack of experimental evidence on how this treatment worked bothered a number of clinicians at the time. The main issue was the lack of scientific proof that the ligation of this artery increased blood flow to the heart. At the same time, a prominent scientist and anesthesiologist from Harvard Medical School, Henry K. Beecher, published an influential paper titled, "The Powerful Placebo".[1] This clinician-scientist wrote this paper based on his observations during World War II, to the effect that a saline solution could mimic the powerful analgesic effects of morphine, when the patients believed that it was morphine. In this article, he discussed the potential impact of a sugar pill on health, especially to alleviate pain. A group of scientists decided to test whether this simple surgical approach (of ligating the mammary artery) effectively treated heart disease.[2] They compared this approach with a placebo therapy.

Five researchers from the departments of Medicine and Surgery at the University of Washington School of Medicine decided to investigate whether the ligation of the mammary artery would effectively decrease heart pain when compared to a sham (i.e., placebo) surgery.[2] They randomized 17 patients to undergo the real or sham (in this case, anesthesia would be administered, and a cut would be made on the skin alone, and nothing more) surgery. The results were very interesting. There were no significant differences in the levels of improvement between both groups. For instance, the estimated subjective improvement was 32% in the real group as opposed to 43% in the placebo group.[2] The authors described remarkable improvements in the placebo group, too. They reported that "One patient (case 4), who had been unable to work because of his heart disease, was almost immediately rehabilitated and was able to return to his former occupation. He reported a 100 percent improvement at 6 months and 75 percent improvement after a year. His arteries were not ligated."

These results put an end to this surgery for coronary heart disease. This study also underscored the importance of understanding the role of expectations in modifying and improving symptoms.
Drawing from this, I frame the following question for learning: Does the expectation of learning affect the efficiency of learning itself? I argue that it can indeed. It may seem infeasible and unethical to give students sugar pills and tell them that their learning would be significantly better if they take the pills. If teachers are not going to give their students placebo pills, why is this important? It is important because placebo or the expectation of a positive outcome may be triggered not only by a sugar pill. Our previous expectations of positive

experience will shape the results of that experience. A placebo pill is only one of the means to trigger such an experience.

Can sugar pills make our brains learn more?

Imagine the following situation. You are presented with a test on a computer. You see Chinese characters on screen. There is an algorithm (which is not given to you) on the characters that are associated with greater rewards. Someone comes and tells you that you will receive a safe and non-invasive form of brain stimulation that will make you possibly (or certainly) learn more (but in reality, this is fake). Would you learn more?

This is what four scientists from the University of Gottingen tested in 30 healthy participants.[3] They were exposed to a test in which they were rewarded with money based on their success. An interesting aspect of this experiment was the placebo certainty. The researchers divided the participants into two groups. They told one group that the method of brain stimulation they were about to undergo had been proven to induce beneficial effects on cognition (low uncertainty). The other group was told that this technique was still an experimental approach to enhance learning (high uncertainty).

Both placebo groups had a significant learning effect. Interestingly, the high uncertainty group had an even larger effect. According to the authors, there was a U-shaped relationship between reward certainty and dopamine release. Dopamine is a neurotransmitter associated with the pleasant effects of a reward. The authors hypothesized that this U-shape relationship led both the low and high uncertainty of a reward to induce larger releases of dopamine, which, in turn, helped the participants learn.

Learning is highly associated with our brain state. Effective learning depends on both good material (either a book, a teacher, etc.) and our brain state. This simple experiment of inducing participants to think that they received brain stimulation (when they did not) shows the importance of preparing our brains to learn.

Do you drink coffee before going to bed?

You may have been involved in small talk of the following kind. You are having dinner with your friends, and after dinner, the waiter comes and says: “Would you like coffee?” Some of your friends say: “Yes, please,” while another friend says, “I cannot drink coffee after 4 pm. If I do, I cannot sleep.” This comment triggers a debate, and about half say that they experience no changes in their sleep patterns, regardless of when they drink coffee, and the other half counters this view passionately. Although there is a simple potential scientific explanation here (which is that some people are more sensitive to the effects of coffee while others are not), this may also be related to a placebo effect. Suppose you drink coffee one night and have a hard time sleeping (not associated with the coffee). What does your brain do? It looks for correlations and finds one hanging there, and decides that coffee made it difficult for you to sleep. Next time you drink coffee after dinner, your brain will pick up that information and process it consciously or (even worse) unconsciously, and now you are conditioned to the idea that coffee at night leads to no sleep. Although I do not dispute the potential differences among different individuals in response to the effects of coffee, a placebo effect may just be the cause.

This is what researchers from the University of Connecticut found (Kirsch and Weixel, 1988[4]). In this study, the subjects received decaffeinated coffee but were told that they were receiving regular coffee. They compared the results against subjects who were told that they may or may not receive real coffee (in a double-blind manner). The results were unequivocal. When the subjects believed that they received real coffee (even if it was not), they had nervous system reactions that mimicked their reactions to real coffee. They had higher pulse rates and different levels of alertness, tension, and systolic blood pressure. This shows that our beliefs or expectations are potent modulators of our brain states. This can be translated into learning and forms the core of the discussion in this chapter. Our expectations of learning influence the learning outcome.

Our personality trait influences the "placebo pill" in education: Horse serum to treat cancer

There is a classic placebo story of a patient dying with cancer that goes back to 1957.[5] His tumor was large, and he was in a critical condition. He read in the news that horse serum was capable of treating cancer. He insisted that his physician, Dr West should administer him the treatment. Dr. West agreed. After the first injection, he improved significantly. It was a miraculous recovery. However, one day, the subject read another news story which reported that the horse serum treatment was fake and thus ineffective as a remedy against cancer. He worsened right after that and became very ill again. Dr. West told him about a new version of the drug, which had an improvement over the earlier one, and assured him that it truly worked. The patient improved again. It was like his tumor mass had melted. This time, he only received water. Finally, he read a report that strongly affirmed that the drug did not work. He was dead two days after reading this final report on this potential drug.[5]

This story tells us that it is not only the expectation that something will make changes in our brains or bodies, but also our belief systems that will determine an outcome. Making the parallel in education, the belief system that is important for the placebo effect in medicine can be translated to the idea of the classical fixed vs. growth mindset in education.

The main idea is simple. In the fixed mindset, students think that their intelligence, learning curve, skills, and abilities are fixed traits that cannot be modified. In the growth mindset, students see learning and intelligence as traits that can be modified and improved by working and studying. It is difficult to categorize someone in either a fully fixed or growth mindset exclusively. Students are often situated in a continuum between these two extremes. Therefore even if a student is exposed to a positive environment that gives him the message that learning will occur, a fixed mindset may not be affected by these placebo-like effects. In other words, a fixed personality trait may be less affected by a placebo conditioning than by a growth mindset trait.

A one-hundred or two-hundred page book?

You are enrolled in a math course. Suppose you have to choose between a book with one hundred pages and another with two hundred pages, as the course textbook and the only source material to study from, for the course. Suppose now that the content is similar (more pages does not mean more content). Which one do you think you would learn more from? The first noticeable and intuitive answer is the longer book. But remember that the content is similar between the two versions. One can argue that more time will be spent studying the topic in the book with more pages. That is true. But longer study material also gives us more massive placebo stimulus.

This is what a previous study found: two pills as opposed to one pill induced a much larger placebo response in healthy subjects that participated in a research study. The study analyzed several clinical trials for gastric ulcer with a placebo arm. The scientists found that four pills of placebo per day had a larger effect than two pills of placebo.[6] This result was maintained when there was an adjustment for patients' characteristic.[6]

Although it is difficult to detangle the hours students spend on studying material and the higher perception and expectation of learning, placebo-like effects in education are likely affected by the perception of the learning dosage. I do not suggest that teachers should give hundreds of pages or assignments and reading material and double the lecture time to enhance learning. However, here, longitudinal course designs may help the learning dosage.

Instead of offering the entire material over two weeks, offering it over two months may have a greater impact. Here, again, there are several factors that play into the drivers of learning and more time and/or exposure to material can have an impact on learning outcomes. However, my point here is that longer exposure can increase one's individual perception of learning and can help with a placebo-like effect in learning.

The Mercedes-Benz effect in education

Another placebo study presented an interesting factor that can mediate placebo effects: branding. In a well-designed study published almost 40 years ago, two scientists from the UK randomized 835 women with frequent headaches to receive placebo or active analgesic drugs.[7] The placebo could mimic an unbranded or branded analgesic. The group that was administered the active analgesic was divided further into two groups, which received either the branded or unbranded analgesic. There were four groups in all: unbranded and branded placebo and analgesic groups. These two scientists confirmed their hypothesis: the branded placebo and analgesic drug induced more pain relief than did the unbranded ones.

These results make us think of whether there is a dose for the expectation effect in learning. Let us think about the following experiment. There are two courses that are exactly the same. They have the same material, instructors, didactics, and students. You are taking one of them at Harvard, and the other at a small community college located in the countryside in a rural state. Would the results be the same? This is almost impossible to conduct, so we can only discuss the potential effects. Although it does seem reasonable to suppose that there will be different effects (again remember that the two courses are identical, the only difference is the university that is offering), it is difficult to understand the factors that predict potential differences in learning outcomes. Is there a placebo-like effect of taking a course at Harvard (that we expect we will learn more), or is this because students at Harvard want to show they can be successful in getting a good score and work harder? Although there is no answer to this question, both potential mechanisms are correlated. One is more associated with the expectation, while the second is more associated with motivation. One important consideration is that expectation can also reinforce motivation. For instance, the expectancy of learning something may make you more motivated to learn. The contrary is also true. Motivation to learn something will increase our expectation for such an experience. Therefore, the placebo-like effects in education may also end up enhancing motivation.

Learning to be a poor learner

You are invited to participate in an experiment. You are told that an electric device will be hooked to your hand (that will not cause any sensation, as it is off), and this will make you more sensitive to pain. You agree to participate. In the next phase, called the conditioning phase, the experiments make you learn that every time they say that this device is ON (again, this is a placebo device and is always off), they deliver a warmer thermal stimulus with another device (thus, this will generate some pain). When they tell you that the electric device is off, you feel the thermal stimulus weaken (less painful) as the investigators intentionally decrease the thermal sensation. In sum, you are told that the thermal stimuli (the painful one) always have the same intensity. But, in fact, they are being adjusted to the ON/OFF status of the device "modulating pain." After several trials, you learn that when the electrical device is ON, you feel more thermal pain.

Next, the researchers tell you that the device is ON, but this time, they do not increase the thermal stimulus (they deliver low thermal intensity). As you learned that the device causes more pain, you do feel more pain even with lower thermal intensity. This is the nocebo effect. This is what a group of researchers from the Netherlands showed in their experiment.[8]

This is an essential concept in placebo research. The placebo effect can have detrimental or beneficial effects. When beneficial, it is called the placebo effect and, when detrimental, it is called the nocebo effect.

Is it possible that if your historically low grades in math led you to believe that you cannot learn math, and that in the next math course, you will fail? The answer may be yes. However, it is not as simple. The nocebo effect can decrease your motivation toward the educational program as even with extra effort you believe there will be no improvement in your learning.

In the example of the college course mentioned above, perhaps it is not that students will feel that they will learn more at Harvard, but that instead, they may feel that they will learn less at the community college course. It is also possible that they may feel both.

In medicine, as mentioned above, this effect is called the nocebo or lessebo effect. It means that when you take a drug or undergo a treatment that you have particular expectations of worsening, it will make you worse in a specific aspect. For instance, you may think that the drug will make you sleepy, give you headaches, and so on, and then you experience that symptom. Even if this treatment is a sugar pill, this is the nocebo effect. The lessebo effect is similar, but the difference is that it removes the real drug's therapeutic benefits. For instance, you take an active drug to decrease pain, but you think this is a placebo and has no effect. You then annul the effects of the active drug by thinking that it is a sugar pill. Scientists do not entirely understand the basis of the nocebo or lessebo mechanisms. It may be related to cognitive changes that can orient our attention system to be more aware of symptoms related to the disease, or may also modulate the neuroendocrine system that may generate other objective changes in our immune and endocrine systems.

Here, it is important to discuss two possibilities for the nocebo mechanism in order to understand its effects on learning. The first is its increase in attention (in the case of nocebo to orienting attention to negative aspects), and the second is its effect on changing neural circuits. The idea that something can make us sick changes our brain activity, which results in changes in our behavior, as well as in our immune and endocrine systems. It is logical that they would also affect learning, as learning involves neuronal activation.

If you think participating in a course would not affect your learning, when you do participate in the course, you may wind up learning less, given that your attention system will not be particularly engaged, or that the learning networks may not be properly activated.

In summary, the previous belief and expectation that students have toward a course or a program and their learning abilities will play a significant role in their learning process. This is not easy to quantify, but it is certainly a factor that many teachers do not look at.

References for chapter 5

1. Beecher, H. K. (1955). THE POWERFUL PLACEBO. *Journal of the American Medical Association*, *159*(17), 1602. https://doi.org/10.1001/jama.1955.02960340022006

2. Cobb, L. A., Thomas, G. I., Dillard, D. H., Merendino, K. A., & Bruce, R. A. (1959). An Evaluation of Internal-Mammary-Artery Ligation by a Double-Blind Technic. *New England Journal of Medicine*, *260*(22), 1115–1118. https://doi.org/10.1056/nejm195905282602204
3. Turi, Z., Mittner, M., Paulus, W., & Antal, A. (2017). Placebo Intervention Enhances Reward Learning in Healthy Individuals. *Scientific Reports*, *7*(1). https://doi.org/10.1038/srep41028
4. Kirsch, I., & Weixel, L. J. (1988). Double-blind versus deceptive administration of a placebo. *Behavioral Neuroscience*, *102*(2), 319–323. https://doi.org/10.1037/0735-7044.102.2.319
5. Blakeslee, S. (1998, October 13). *Placebos Prove So Powerful Even Experts Are Surprised; New Studies Explore the Brain's Triumph Over Reality*. The New York Times. https://www.nytimes.com/1998/10/13/science/placebos-prove-so-powerful-even-experts-are-surprised-new-studies-explore-brain.html

6. De Craen, A. J. M., Moerman, D. E., Heisterkamp, S. H., Tytgat, G. N. J., Tijssen, J. G. P., & Kleijnen, J. (1999). Placebo effect in the treatment of duodenal ulcer. *British Journal of Clinical Pharmacology*, *48*(6), 853–860. https://doi.org/10.1046/j.1365-2125.1999.00094.x
7. Branthwaite, A., & Cooper, P. (1981). Analgesic effects of branding in treatment of headaches. *BMJ*, *282*(6276), 1576–1578. https://doi.org/10.1136/bmj.282.6276.1576
8. Thomaidou, M. A., Veldhuijzen, D. S., Peerdeman, K. J., Wiebing, N. Z. S., Blythe, J. S., & Evers, A. W. M. (2020). Learning mechanisms in nocebo hyperalgesia: the role of conditioning and extinction processes. *Pain*, *161*(7), 1597–1608. https://doi.org/10.1097/j.pain.0000000000001861

Chapter 6: Does Self-confidence Play a Role in Learning? The "Big-Fish Little-Pond" Effect

In March 2019, prosecutors from the Office of the U.S. Attorney for the District of Massachusetts revealed a bribery scheme case involving parents who attempted to have their children admitted to the top prestigious colleges in the U.S., including Harvard University, Stanford University, the University of California Los Angeles, and the University of Southern California. Among the 33 parents charged in the case, two names stood out due to their roles in the entertainment industry: the Academy Award–nominated actress Felicity Huffman and the *Full House* series actress Lori Loughlin. In Felicity Huffman's case, she hired someone to take the SAT for her daughter so that she would obtain better scores in her college application. Lori Loughlin and her husband went one step further through bribery—they bribed college personnel for their daughters to be admitted as recruits in the crew team, even though they never participated in the team. Both

Lori Loughlin and her husband ultimately pleaded guilty and served time in prison.

This was a highly publicized case that dominated the news cycle for several weeks (even with the emergence of the COVID-19 pandemic). Most news stories focused on wealthy parents trying to circumvent the system with their wealth. However, this is not the point I want to discuss here. There is another side to this news story that is vital to consider, and I have not seen any discussion on it thus far: were these parents *helping* or *hindering* their children? If the saying that it is better "to be a big fish in a small pond" is true, then this bribery case would imply that these parents were likely placing their less prepared children in a pond that was too large for them. In fact, the associated risks and consequences of these parents' actions could ultimately result in a negative learning outcome for their children. It could be argued that if they did not commit their illegal actions, their children would have been better placed as big fish in smaller, less competitive colleges.

Given this notion, this chapter discusses that our remedy for someone with a weak academic performance (i.e., placing this person in a top school) is not necessarily the appropriate medicine. As the Latin quotation states, “In regions caecorum rex est luscus”—“In the land of the blind, the one-eyed man is king”. In brief, this statement can be interpreted as a skill becoming an advantage compared to a person’s weaker surroundings. In this chapter, I discuss that the likely reason for that is that it would increase self-confidence and self-efficacy. In light of this idea, an important question can be asked: should we all follow our unconscious minds and desire Harvard and Stanford for our children or us regardless of our self-efficacy?

The “Michael Jordan” effect: The importance of the self-concept

Michael Jordan is often considered the greatest basketball player of all time. He has helped the Chicago Bulls win six National Basketball Association championships and has had several MVP seasons. He is undoubtedly an outstanding learner of the sport. However, we cannot ignore that he had certain characteristics that helped him achieve this high place in the game’s history; he was not born with the basketball skills for which he is known. He had to work hard and learn to be outstanding, and this is something that he has acknowledged himself. One of his famous statements was: "I've missed more than 9,000 shots in my career. I've lost almost 300 games. Twenty-six times I've been trusted to take the game-winning shot and missed. I've failed over and over and over again in my life. And that is why I succeed". With this claim, Michael Jordan is revealing a well-known secret: learning is about trying in a focused and strategic manner. This is why the mindset is crucial. If Michael Jordan felt like he was a failure after every mistake he made, then he would not have had the energy and motivation to continue trying and finding his extraordinary playing style.

One important reason why Michael Jordan continued to develop is that he maintained a high self-concept, which is similar to self-efficacy. Both terms denote the belief that people can succeed in any given task and maintain their success in the long term. Michael Jordan certainly believes in his skills to succeed and continue training, trying, and improving. Of course, his natural physical and mental characteristics were fundamental to his success, but they were far from being the only reason. Put another way, his natural abilities helped him develop his self-concept, which in turn helped him become an effective learner.

This is a key factor of learning: how we feel about it. If learning is an enjoyable experience, then we will want more. Human beings crave positive experiences. If a 10-year old is the best student in his class, then he would feel proud of himself; he would enjoy his parents, peers, and teachers' compliments. He would want to do more—he would know that he could do more. Positive reward is one of the best fuels for learning. Conversely, for students who fail and experience negative reactions, learning becomes a painful experience.

In a personal example from my life, I remember how my middle school teacher used to assign simple mathematical exercises to the class. The first student to answer had to raise his or her hand, and I still remember the positive feeling of answering more quickly than my fellow students. However, I did not want to do well in math so that I could become a mathematician or engineer; I simply enjoyed the subject, as well as reading books with mathematical puzzles that my father had given me. My passion for mathematics and the positive feedback I received had a reciprocal effect on my performance. Considering how Michael Jordan learned to excel at basketball and how I learned to do well in mathematics, it is crucial to question how self-concept truly contributes to the learning process.

The Framingham cohort and the idea of effect size

Framingham, a suburban city in west Boston, has become particularly famous. This town is the origin point of one of the most famous medical studies: the Framingham cohort. Promoted after President Franklin D. Roosevelt's premature death from hypertension and stroke, this study determined the critical risk factors for heart disease. It was vital to establishing risk factors and determining their effect size for cardiovascular disease. In short, we not only learned that hypertension causes heart disease but also that this relationship had an effect size. Mild hypertension leads to heart disease, but the effect size is relatively small. That is, mild hypertension will, on average, have a mild-to-moderate effect on cardiac function. This story about this landmark study in medicine is to highlight the important of effect size. Effect size explains how meaningful the relationship is between variables, or how large the difference between groups is. It represents the practical significance of a given research outcome, and it is an essential notion to consider when discussing the big-fish little-pond effect in the context of self-concept

A group of education scientists in the United Arab Emirates investigated the effect sizes of the relationship between self-concept and mathematical achievement.[1] These researchers analyzed data from 7,404 adolescents from 456 schools in the United Arab Emirates. They measured the students' self-concept in mathematics using Likert-type questions that ranged from "strongly disagree" to "strongly agree"—such as, "I am just not good at mathematics," "I get good grades in mathematics," "I learn mathematics quickly," "I have always believed that mathematics is one of my best subjects," and "in my mathematics class, I understand even the most difficult work." The researchers measured mathematical performance using a composite assessment that contained standard multiple-choice questions, complex multiple-choice questions, and constructed response items. Two critical findings were discovered in this study, of which the first was the relationship between mathematical achievement and mathematical self-concept, with an effect size of 0.64 (which is considered a large effect size). The second finding was the negative correlation between school average mathematical achievement and mathematical self-concept. This indicates that a higher school average is associated with lower mathematical self-concept in any given student.[1]

This study also provided crucial data that demonstrated the relationship between students' performance and their self-confidence. It found that the better the student group and school, the lower the self-confidence levels of individual students. In light of this finding, the study contrasts the parents' desperate and illegal actions to enroll their children into ivy-league schools, as discussed in the opening of this chapter. The United Arab Emirates study demonstrated that both talented students and academically disadvantaged students could have decreased self-concept in these top schools.

However, two important questions must be asked at this point: Is the relationship between school averages and self-concept uniform over age differences—that is, is this relationship's effect on adolescents equal to its effect on college students? Moreover, is the relationship between self-concept and performance causal?

Let us tackle the first question: age and the big-fish little-pond effect. A study from a group of researchers in China analyzed 33 studies (total number of students = 1,276,838) and revealed that age is indeed an important factor that mediates the big-fish little-pond effect.[2] In fact, the largest effect was observed in high school students, and the smallest effect was detected among primary school students. Students in middle school and college demonstrated moderate effects. This analysis is intuitive, as high school students are in the age range that experiences the strongest effects of social and peer pressure.[2]

The second question mentioned earlier can be further refined to ask whether low achievement leads to low self-concept or vice versa. This strongly concerns the causal relationship between the big-fish little-pond effect and academic achievement. Although a relationship between self-concept and academic achievement clearly exists, it is unclear whether this relationship is causal, or whether it ensures that low self-concept results in low academic achievement (and not vice versa).

Although self-concept evidently decreases as the school's average achievement increases, this does not sufficiently affirm that low self-concept leads to low academic performance.

Concerning the studies and findings discussed above, my view is that a multivariable relationship defines this relationship—and that self-concept is one variable. Its effect also depends on students' support systems and other emotional variables to which they are exposed.

After this discussion on effect sizes and academic achievement, I have asked myself whether I would avoid sending my children to a top school. In truth, I would not; however, I would be aware of this big-fish little-pond effect, and I would try to counterbalance it. Most importantly, I would not prevent my children from attending a school of average academic achievement if it was their choice, as this choice could ultimately prove to be an advantage in their overall learning. Finally I would focus more on my children self-concept rather than the US News Best Colleges ranking.

How to reduce the "little-fish large-pond" effect

So far, this chapter has discussed how the big-fish little-pond concept affects students' ability to learn. Although this effect is mild to moderate—indicating that it contributes to learning, though it is not the only determinant of learning—it is still an important factor in education that must be studied. This effect is especially important for high school students. As such, we must understand how we can decrease this effect when students are little fish, or we must determine whether it is possible (and desirable) to enhance their self-concept, even when they are large fish.

One main factor driving the effects of self-concept are the wrong weights that are attributed to the outcome (grading) and process of learning. As discussed in other chapters, students attribute too much weight to grading, and they consider it the metric of their learning success or failure. Understandably, students would lose motivation after receiving low grades, as this is their metric of success. Notably, Asian students are seemingly more affected by the big-fish-little-pond effect. Asian families, on average, attribute considerably more weight to grading as a metric of success in life.

To decrease the effects of surrounding academic achievement on an individual's self-concept, the emphasis on grading must be reduced—and this should be implemented for all actors involved (students, teachers, and parents). First, teachers should change the criteria for successful/passing grading. Instead of taking exams, students should be assessed according to their time engaged in learning tasks. This solution is not easy to implement, and it does not completely eliminate the problem, but it can improve it. Second, students and parents must understand that grading is not a measurement of intelligence or learning capacity; they should learn that the metrics of success should be multifactorial.

Olympic world records and competition in education

One aspect of the Olympic Games that most spectators enjoy is when a world record is broken. Winning a gold medal in the Olympic Games is more valued when it is accompanied by the breaking of the world record. In a way, this is an extreme manifestation of the big-fish little-pond effect. Olympic competitors want to be the best not only in the pond of the current Olympic games but also in the entire pond (all previous competitions). This drives athletes to be the ultimate best and keeps encouraging world records to be broken.

At this point, we can return to Felicity Huffman and Lori Loughlin from the beginning of this chapter. Did their illegal actions at least help their children or push them to excel? Suppose we conclude that highly competitive environments can push performances to extend beyond previous records. In this case, it could be argued that a highly competitive environment is one method for pushing students to become extraordinary. However, this is not the case; it certainly did not help Huffman's and Loughlin's children. The important difference to note in this case is that athletes aiming for the Olympics start their journeys by being big fish in their little ponds; they are the best in their countries, so they have a high baseline self-concept. These athletes are also mentally and physically prepared well in advance for the Olympics, and the competition lasts only for a few days; their self-concept cannot significantly decrease in such a short amount of time.

Let us consider as an example the men's final 100-meter event in the Rio 2016 Olympic Games. Usain Bolt won the final match with an impressive time of 9.81 seconds, while Trayvon Brommel from the U.S. finished last with an impressive time of 10.06 seconds. Now, imagine that the athletes from this race kept training together for several years, and Brommel continuously ended last in the race. Regarding Brommel's self-concept, it is easy to guess that he will likely lose his motivation to continue learning. However, if he were competing in the 1980 Olympic Games, he would have been the fastest runner in history, given his impressive time of 10.06 seconds (the fastest runner in the 1980 Olympic race was Allan Wells from Great Britain, with a time of 10.25 seconds). This example reveals that a person's current surroundings, or pond, play a significant role in their self-concept and desire to learn and improve. But this example does show that it can work in both ways: to increase or decrease motivation. The key, therefore, is to understand the unconscious thinking associated with surroundings (e.g., pond) deliberately.

In the context of competition and education, the secret is to focus on improving and shutting down surroundings. Students should focus on their learning trajectory and avoid as much as possible comparing with others learning trajectories. This is the basis of interventions focusing on meta-learning.

Learning to turn distractions off

In the movie *Love of the Game*, Kevin Costner plays a baseball pitcher for the Detroit Tigers. In one scene, Costner pitches against the Yankees in their stadium, which is filled with noise from Yankees supporters. He activates this perfect focusing mechanism and throws a perfect ball (this scene can be seen on YouTube: https://www.youtube.com/watch?v=ykV1z7NPr60). Although it is just a scene from a movie, it illustrates the importance of having a perfect focusing mechanism. This is important not only for playing baseball but also for modulating the motivation system.

The key to modulating our motivational system is knowing how to feed on it. If we constantly feed our motivation system with information like "I can't make it," "I always fail," and "my colleagues can do it and I can't," then we signal that we are experiencing something painful—and our reward system consequently wants us to draw closer to what promotes positive feelings rather than to what promotes pain. In this sense, we need an active cognitive analysis of our surroundings. Instead of comparing ourselves to others, we should enjoy every victory, no matter how small, and feed the motivational system accordingly. Our focus must be directed toward gains and not toward relative losses. As Michal Jordan has stated in several interviews, he habitually focused on specific goals—even on one small goal each time—and he ignored the cheering from the audience, overall scores, and other distractions.

Focusing on gains helps our motivational system concentrate on the task at hand, which prompts us to learn more. However, the concept of loss dominates our thinking and creates noise. In fact, the scene with Costner from *Love of the Game* reveals a crucial feature of our brain: we need a strong inhibitory component to harmonically process all the information we absorb. Information overflow does not only occur in baseball finals; it occurs constantly through internal and external stimuli. If these stimuli are not inhibited or controlled, then there is no meaningful brain activity.

One important aspect to note here is that the ability to focus demands brain energy. When not in focus, our brains return to their default survival mode (which usually amplifies negative threats). We have less energy for brain inhibition when we are tired or hungry, and thus less energy to control the surrounding effects and focus on what matters. This is why teachers, students, and parents are critical; they help others focus their attention on learning gains rather than on surroundings. As discussed throughout this book, being the first in class is not what matters, and it is not what makes someone successful. Success is determined by how well people can learn in a certain domain, and how they can utilize their learning to benefit society.

"Big-fish little-pond:" Effect on older-aged people

In the process of aging, neurodegeneration (especially that of the frontal lobe) affects vital brain components that relate to inhibitory control, which leads to a lack of focus. Although we would expect less social pressure on older people due to the development of other psychological mechanisms of compensation, we should still consider the self-concept effect that is associated with learning. In this case, older-aged people might be more difficult to modulate the reward system, given the lack of a strong inhibitory system. Additionally, older people can experience negative learning, which occurs when adapting to a loss of sensorial input (meaning-related information obtained from sensory systems) leads to worsening performance due to the lack of inhibitory control.

The idiomatic phrase of "use or lose it" applies to brain function, especially to our brain inhibitory system. In this sense, switch-off mechanisms (inhibition) are necessary for us to achieve balance across our 80 billion neurons, as well as vital for us to develop our intellectual skills and learning. As discussed in this chapter, throughout our lives, we need to develop our learning strategies to improve our overall learning and keep this system active and strong to focus on what is important and not on the surroundings.

References for chapter 6

1. Areepattamannil, S., Khine, M. S., & Al Nuaimi, S. (2017). The big-fish-little-pond effect on mathematics self-concept: Evidence from the United Arab Emirates. *Journal of Adolescence*, *59*, 148–154. https://doi.org/10.1016/j.adolescence.2017.06.005
2. Fang, J., Huang, X., Zhang, M., Huang, F., Li, Z., & Yuan, Q. (2018). The Big-Fish-Little-Pond Effect on Academic Self-Concept: A Meta-Analysis. *Frontiers in Psychology*, *9*. https://doi.org/10.3389/fpsyg.2018.01569

Chapter 7 - Enhancing Learning: Metacognition as Low-Hanging Fruit?

*"The definition of insanity is **doing the same thing** over and over again, but **expecting different results**." (usually attributed to Albert Einstein*[1]*)*

The 1984 Monaco Formula 1 race presents an extraordinary example of overcoming challenges by doing things differently. On the day in question, Monaco had heavy rain, so much so that the start time of the race was delayed by 45 minutes. Ultimately, the event had to begin during the downpour. Conditions are dramatically different under the rain, and existing advantages disappear if competitors are not well trained for such circumstances. Accordingly, these conditions were challenging for most of the drivers. They were, however, a blessing for Ayrton Senna, a young man on an unquestionably inferior Toleman team. Notably, this team's car did not match the quality or sophistication of the McLaren racecars, either. But Ayrton knew how to drive in the rain. He had trained for that possibility, and this gave him a significant advantage.

It is noteworthy that his motivation to learn this specific skill came from earlier racing defeats under wet conditions. This demonstrates an important aspect of Ayrton's mindset that we will discuss in this chapter: his metacognition skills. Often, when we face defeat, our immediate thought is that we did our best, and it simply did not work despite it. We may also start to think, "This is it for me. I will never make it." Yet, we also have the option to analyze the situation and change our actions. When Ayrton was defeated in the past, he could have given up or continued training in the same way he did before. The result of this would be that he would continue to lose. Instead, he decided to assess the reasons for his defeat, along with his weaknesses, and consider how he could improve. It seemed that his difficulty in performing under wet conditions was a weakness. He then turned this disadvantage into one of his greatest strengths, becoming a winner under the conditions he previously struggled with. Ayrton Senna became one of the most prominent drivers in the history of Formula 1.

Ayrton Senna's approach is a good example of metacognition: assessing the learning process to allocate resources appropriately. Moreover, this act of continuously assessing his learning strategy was a key factor in his success. Broadly, metacognition refers to thinking about the thinking process. This means not only engaging in the process of thinking (or learning, or studying) but also pondering over the thinking process itself. Accordingly, we will discuss some helpful strategies to break the patterns discussed in the previous chapters.

The powerful force of experiential learning: squirrels and bird food

In our previous home in Chestnut Hill, Massachusetts, my wife and I always enjoyed seeing birds come into our backyard. Accordingly, we decided to feed them wild bird food. In doing so, however, we failed to account for a major obstacle: squirrels. As soon as we put the food out, they quickly came and ate all the seeds. We then hung a bird feeder from a tree branch, believing we would protect the food from these clever creatures. This only worked for a few hours, until they learned how to lean on the branches and access the food.

I did not give up. I bought a tall pole that hung the feeder at the very top, thinking I had finally won the battle. Again, the solution did not last long. The squirrels mounted our fence and—to our astonishment—learned to jump to the bird feeder and devour the food. I retreated from the fence. They, again, defeated me. In my last attempt, I sprayed the pole with olive oil. This was somewhat successful, but some learned how to climb the pole fast enough to overcome its slipperiness. "Oh well," we thought. We gave up and decided to buy more food for both the birds and squirrels.

Why am I telling you this story? It exemplifies experiential learning, that is, learning through experience. This term and the theory behind it were developed by the psychologist, David Kolb.[2] The concept is simple but formal classroom education has taken students far from this natural way of learning. Here, the term "natural" is used because (much like my bird feeding saga) experiential learning is learning through experience—in other words, by trial and error. Indeed, experience is essential, as this kind of learning requires that one reflect on a previous experience. After the initial undertaking, the student must reflect on what went wrong (and right) to change and adapt their learning strategies. These are the basic principles of Kolb's cycle, in which a student is exposed to the experience and then reflects on it to adjust his action and learning strategy.

How does this scenario differ from that of a student attending and learning at a traditional school? Usually, and unfortunately, they are fundamentally dissimilar. In the school environment, the student is not exposed to the experience first. Instead, a student will be told that squirrels are notably intelligent animals that will learn to jump, climb poles, etc. In other words, they are given a response before they can even think of a question. Yet, this approach provides only abstract learning and is likely to be quickly forgotten. Conversely, I will never forget my day-to-day battle to feed my bird friends.

Using metacognition skills: unlimited benchmarks for success

Another problem in the traditional school system is the assumption that a perfect grade (or the equivalent of 100% / A+) is the benchmark for total success in learning. There could not be an approach less conducive to metacognition learning strategies. There is always room to learn more and to continuously assess different strategies to enhance learning. Conversely, a student that uses a grading benchmark as their guide will likely become complacent and decrease their learning capacity; they will not change their strategy to learn more about the subject.

Moreover, to a student led by grades, a low grade leads them to believe that they are not "intelligent" or "capable" when this is not the truth. Depending on how the grading process is defined, it may only reflect a student's ability to respond to what the teacher wants. And yet, a learner's goal is not to learn by giving the teacher the answer he/she wants but to learn comprehensively, which will also facilitate the learner's career goals. One should never stop learning. An A+ or a diploma does not mean that the learning is complete. This is only a standard that a student of traditional education needs to fulfill. Unfortunately, very few make this distinction.

Providing analgesics that lead to more pain: learning assessment and metacognition

One common oversight doctors acknowledge is that the lack of time to listen to patients' complaints and medical history in their entirety usually leads to prescription errors. Tension headaches are a common pain syndrome. Studies show that approximately 50% of the population will suffer this ailment at least once in their lifetime.[3] A pain doctor who sees a patient with a tension headache may dismissively think, "Oh well; this is not a difficult case, I will prescribe a common pain killer." When the patient returns in a few weeks, complaining of more headaches, it becomes clear that the initial assessment may have been wrong. The patient's pain was induced by chronic analgesic use, that is, the excessive use of analgesics caused even more pain. Thus, the physician worsened this situation with their quick assessment and generic prescription.

Learners also frequently struggle with improper diagnosis. Usually, teachers give more "analgesics" when a student is falling behind, for example, providing more readings and study materials for the pupil in question. Yet, the problem is rarely the lack of material, but the lack of resources used to study it. For instance, students may lack intrinsic motivation or focus because they have minimal interest in the material (that may be enhanced by the factors discussed in the previous chapters). Moreover, the educational program may already be overwhelming the student with content (causing cognitive overload) or the learner may not have an incentive to engage with these resources. As in the headache example, administering more analgesics will only worsen the problem.

The first step toward resolution requires students to learn about themselves. If they do not like the subject, that is the first thing they need to work on. Indeed, simply forcing oneself to read will only make one dislike the subject more.

How can the diagnosis be made? There is no special exam or machine that will do this for you. However, metacognition fulfills this need. The student must also be an observer of their own learning process. They must be able to see what works and does not work for them and learn how to harness the approaches that foster learning. They also need to assess the unconscious forces (discussed in the previous chapters) that may be taking them away from their effective learning strategies.

The Moses illusion and the need for insight into metacognition

There is a classic cognitive psychology test called the Moses illusion. In it, participants were asked, "How many animals of each kind did Moses take on the Ark?" Usually, people correctly respond with two, not even noticing the inconsistency at the heart of the question. It was Noah, and not Moses, who took the animals to the Ark (Reder, book chapter). There are several theories to explain why this distortion is not detected, but all seem to point to a common mechanism: our cognitive processes favor efficiency over accuracy.

When this question was asked, the individuals were thinking about the right answer to the question instead of checking the accuracy of the question itself. This oversight was worsened by the semantically similar terms used (Noah and Moses are both important biblical names). If individuals were asked to detect the inconsistency of the sentence or even asked about Moses taking the animals to the ark, their responses would have been different.

The Moses illusion highlights an important characteristic of human cognition: efficiency leads to error. Moreover, the only way to correct that shortcoming is to first gain insight into it. Learners need to understand the basic mechanisms of human cognition processes as a first step toward truly grasping and self-diagnosing their ability to learn.

The accident of Varig Flight 254 and the measurements to adjust your learning strategy

On September 3, 1989, Varig (a now-defunct Brazilian Airline company) flight number 254, departed from Sao Paulo *en route* to Belém (a city in the Para province near the end of the Amazon forest in Northern Brazil). This trip had several layovers and was the penultimate flight available from the city of Maraba, also located in Para. This last leg from Maraba to Belém was relatively short—approximately 215 miles. Before departing, the Captain adjusted the flight parameters according to the flight plan, inputting 0270. Yet, the correct value, in this case, was 027.0. Technology was limited during this time (1989), especially in the remote areas of Brazil. Thus, the mistake made by the Captain sent the plane in the opposite direction to that he intended. As the expected flight duration came to a close, the Captain contacted Belém airport for landing clearance, but personnel could not make a visual detection. Belém is a large city, yet everything appeared pitch black, so much so that airport staff even asked if there was a power shortage in the area. They only noticed the mistake (use of 0270 instead of 027.0) after 30 minutes of trying to find the city or airport. At that point, they tried to locate another close airport, which proved unsuccessful. Ultimately, they had to perform an emergency landing in the forest. Twelve lives were lost.

This story exemplifies a significant challenge in adjusting our learning strategies: how to correct our learning navigation map in real-time—instead of waiting weeks, months, or years to realize that it does not work. The pilot of Varig flight 254 could not see that he was going in the wrong direction. Indeed, he was passing over a dense jungle, so the scenery looked the same on all sides. To make matters worse, the flight crew was apparently listening to a local soccer game, only adding to the distraction. And perhaps even worse than that, by the time the error was identified, it was already dark. In other words, the mistake could only be detected when there was little time to correct it. The situation was only made direr by the limited communication and lack of sophisticated flight tracking technology and devices.

This raises the question of how we can track our learning if it may take months to see whether it is making a meaningful impact on our lives. If we only realized that we did not learn enough in engineering school when our first bridge construction project fails, it would not only be a monumental error but one we had no time to correct. One important point here is that mistakes are important for learning but specifically when we can critically assess those errors and learn from them. This requires an accurate assessment.

Of course, we can adjust our learning strategy based on our grades. However, as discussed in other chapters, grades are not comprehensive learning assessments. For instance, they do not measure learning that endures. In some cases, there may be a correlation between grading and long-lasting learning, but these are in the minority. As in the pilot's case, relying on a faulty method to alter learning strategy may prevent us from finding the correct route and send us even further from the desired destination.

At this point, you may now expect that I will tel bl you how to find the perfect "learning compass". Unfortunately, there is no such thing. Still, thinking about the learning process is an effective method to help you reach your destination. Moreover, this approach means that the learning process is ongoing and enjoyable. If your study and learning time feels like a painful chore, you are unlikely to experience meaningful learning. Remember that our survival instinct draws us away from pain in favor of pleasure. Thus, one's learning strategies must provide one with a pleasant experience. The key to identifying what works best for you is to observe and analyze as much as possible. Instead of using benchmarks of success defined by formal courses, which are usually useless, pay close attention to your own learning strategies and outcomes. Ultimately, the methods and outcomes that create a genuine, lasting impact in your life are the ones you should rely on.

For instance, it does no good to take a 2-week statistical course, receive an A+, and get a nice certificate, only to fail later on when it is time to use your statistical skills to complete a work task. If you have learned only enough to respond to the questions in your course, and not enough to apply knowledge and skills to your real life, your learning was superficial.

The secret of metacognition: there is no magic pill

There is no big secret to metacognition. All the crucial information is contained in its name. It literally means "thinking about thinking"—something we rarely do. Instead, we are constantly thinking about this issue at hand. In other words, we may be thinking a great deal, but not in an effective way. Though this observation sounds like common sense, it is very significant. Since everyone has just 24 hours in a day, the only way to improve is to become more effective within this limited amount of time.

Though we can get more results this way, we rarely consider this point about learning effectiveness. Instead, we devote little time to thinking about our learning strategies, while spending much time trying to learn the things we want or have to—often using ineffective strategies. Indeed, many of us are likely to have poor learning habits.

For this reason, I introduced a second challenge that serves to change your habits. Not only do I ask you to think about your learning strategies, I ask you to change your bad learning-related habits. But, keep in mind, there is no magic pill. There is no 7-day method that will instantly enable you to learn more. As with all lasting learning, you need to change your habits gradually, while investing time in correcting the bad ones. This investment will not yield significant interest rates—only large gains over time due to the compounding effect.

Would you choose a 1% interest rate investment? The learning compounding effect

In the book *Atomic Habits*, author James Clear makes a good case for small changes over time.[4] He shows that by changing 1% of our habits daily (with a corresponding 1% improvement), at the end of one year, we would have improved exponentially. By contrast, if we get worse by 1% each day, our performance will be close to zero in less than a year. This is the secret and, at the same time, the challenge. This is a challenge because small, incremental changes are barely noticeable, thus there is a lack of incentive to continue. However, if Clear's logic resonates with you, it makes a convincing argument to change a little bit every day.

It also highlights an essential learning problem: we tend to want major results in a short period. We want to go from zero topical knowledge to being well-versed on a subject in a matter of weeks. This is unrealistic. Knowledge is built over time. Moreover, when there is this pressure at play, all the energy is devoted to learning the content and not learning about learning itself.

This becomes a vicious cycle: As we go through our school education, there is literally no effort made to teach us to become better learners. Just imagine if kids could make a 1% daily improvement to their learning strategies and skills from kindergarten onward. We would certainly have much better learning outcomes. Here, one may argue: "but how about maintaining enough time to learn mathematics and English?" Yet, option one does not preclude option two. In fact, we can achieve option two (learning core subjects more efficiently) by becoming better learners using option one. By becoming better learners and using facilities to access resources and information (literally one click away, thanks to the internet), we can then literally learn anything— anytime and anywhere. This is the beauty of metacognition: empowering the learner.

References for chapter 7

1. Wilczek, F. Q. M. (2015, September 23). *Einstein's Parable of Quantum Insanity.* Scientific American.

https://www.scientificamerican.com/article/einstein-s-parable-of-quantum-insanity/#:%7E:text=%E2%80%9CInsanity%20is%20doing%20the%20same,usually%20attributed%20to%20Albert%20Einstein
2. Kolb, D. (2014). *Experiential Learning: Experience as the Source of Learning and Development* (2nd ed.). Pearson FT Press.

Appendix 1 Chapter: Research Studies on Meta-Learning: Does Research Support Meta-Learning Strategies to Enhance Learning?

Guest author for this chapter: Anna Marduy

In the last chapter, I invited one of my colleagues Anna Marduy to discuss with some research findings testing techniques to enhance learning by counteracting the inherited systematic errors in our learning trends from our unconscious minds that can decrease our learning efficiency given the design of our educational system.

Before we embark on discussing the research findings, I want to explain why I wanted to have a full chapter to describe and discuss the research studies in this topic for two main reasons. The first one is that this book's main topic is likely new for some readers, and I want to show that this is not only a theoretical problem, but that novel methods to reduce the burden of our unconscious learning tendencies through the use of self-reflection to assess our thought processing, also known as meta-cognition, have been tested in several studies. The second important point is to show that research in this field is not easy because it is truly difficult to measure learning. Moreover, most of the techniques tested were used with regular and traditional education that by itself can elicit some of the biases discussed in this book.

But I see this is a first step to change educational strategies. Indeed, research is critical to change educational policy. One of the main issues that I acknowledge while writing this book is that most of the issues discussed are inherently connected to our current educational system, and to decrease the impact of these biases, we would need to have critical and significant changes in our educational system. In this context, research helps to catalyze such changes.

Finally, these studies discussed below can also help foster new ideas to be applied in future educational programs designed by the readers of this book. These studies are not meant to be the most comprehensive summary of the literature on this topic but rather some examples of well-conducted studies that can also inspire researchers in this area to design and conduct follow-up studies.

The five studies that will be discussed are shown below, and again thank you for my colleague Anna Marduy to have brilliantly analyzed and summarized such studies:

1st study: how early children should be introduced to meta-learning strategies to enhance their learning and whether there is an impact on their learning.
2nd study: The effects of online collaborative learning using forums of discussion to enhance meta-learning
3rd Study: Enhancing online communication to improve learning through self-monitoring and self-regulation
4th Study: Testing Meta-Learning Techniques in Biomedical Sciences
5th Study: Using Flipped Learning to Improve Learning (Measures as Concept Threshold)

1st Study: Children should understand how they learn as early as possible

The concept of meta-learning has been proven beneficial to the construction of lifelong learning (1). Therefore, bringing up the self-awareness of learning from an early age contributes to more successful and efficient learning strategies throughout an individual's life (2). In her article, "Do Ten-year-old Children in Sweden Know How They Learn? A Study of How Young Students Believe They Learn Compared to their Learning Style Preferences", Lena Boström investigates children's perspective on how they think they learn best and compares it to their actual learning style preferences. The study thus seeks to answer the questions *how do young students believe they learn compared to their learning style preferences?* When a teacher learns a student's perception of how they learn and what learning styles work for them, a more student-centered teaching approach can be developed (2).

Boström, thus, qualitatively and quantitatively investigates first-grade students' perception of learning and their learning style preferences and how they compare to their actual learning preferences. For this, the study utilizes three different assessments: 1) the Learning Styles Inventory (LSI); 2) students' verbal estimates of their learning preferences; and 3) interviews with 15 students (2).

The LSI is a questionnaire with 110 statements to which students can respond on a 5-point scale ranging from 1 (definitely disagree) to 5 (definitely agree). Through the LSI, the investigator identified each student's learning style preferences (3). The verbal estimates and interviews were used to evaluate how the students think they learn. Verbal estimates were collected publicly in the students' classroom, and the interviews were done individually with only a small sample of the cohort. Students were asked about their strengths and needs in learning regarding 17 of the 20 elements in Dunn's Learning Styles Model (motivation, conformity, and persistence were excluded).

The questions were related to what was important for the student when they learn. Most of the students said they required a calm working atmosphere; the importance of perceptual and sociological preferences (e.g., movies, books, computers, activities), to which the students broadly answered with a combination of aspects. Most of the students were confused about how they learned, and most referred to taking quizzes or tests when asked about how they learned (2).

It was found that many children overestimated almost all of their strengths, needs, and preferences when it comes to learning (i.e., what senses they learn best with, at what time of day, and in what environment) compared to their actual learning profiles. However, this discrepancy can also be attributed to the preconceptions and opinions children already have about their learning experience and basing their estimates on their classmates' answers, further implying that meta-learning involves a social component (2).

Moreover, in the context of meta-learning, children often found it difficult to express the ways that they learn best in the study. Although this ties into the development of learning and acquisition of cognitive skills, it may also have to do with them not being introduced to meta-learning at an early stage, and therefore lacking awareness in the best way to learn and opening space for the creation of cognitive biases (2). The implementation of educational discussions on learning can help students more clearly describe how they learn, thus leading to more practical learning strategies.

In sum, Boström's study emphasizes the need for the introduction of meta-learning at an early stage in children's academic life. If children understand how they learn, they become better equipped to implement effective learning strategies and bolster their lifelong learning. The exposure of meta-learning to young students should come through their own life experiences and communication with their peers, as these factors deeply influence their learning style preferences.

Therefore, this method could counteract some of the grading effects in guiding (or misguiding) our learning strategies, as discussed in Chapter 2. In addition, the early-on introduction of meta-learning to a student's learning process steers away from the affective impact caused by grading and provides the students with more control over their knowledge retention. Thus, meta-learning provides a different motivational pathway from grading in the learning process, stimulating interests and enjoyment rather than tapping into pressure and fear as motivational aspects for learning (4).

2nd Study: The effects of online collaborative learning using forums of discussion to enhance meta-learning

The use of online-based tools to enhance student learning experience has recently gained traction in academic institutions, with at least 32% of students taking at least one online course (5). The creation of blended learning styles in which online resources are used along with face-to-face interactions has been shown to provide comparable or better outcomes in learning experiences (6). Many blended learning strategies include the use of an online asynchronous discussion board to yield an effective understanding of concepts and collaboration between students (7). Inter-personal collaboration in asynchronous discussion boards builds critical reflection skills on course content and creates an educational Community of Inquiry, allowing the development of organically generated learning patterns (8,9). In their study, Calderon and Sood seek to answer the research question: *What are the organically emerging patterns of learning evident in the content and communication structure of asynchronous discussion board posts?* (10).

Online discussion boards are thought to create independent, self-regulated learning for students, especially in the context of higher education. They promote dialogue and allow students to enhance their critical thinking and communication skills as well as their decision-making process (11). In online discussion boards, students are allowed to post comments, questions, or responses to prompts and reply to other's posts creating a collaborative learning environment known as a Community of Inquiry. A Community of Inquiry enhances pedagogical practice by including three different presences in the learning experience: a social presence, characterized by the interaction between students on the discussion board; a cognitive presence, represented by the students' reflections upon prompts and course content posted on the discussion board; and a teaching presence, characterized by the role of the teacher as a facilitator on the discussion boards (8,9).

Studies show that active interactions between students through discussion boards develop more profound contemplations of a course's content and can be associated with better grades and higher levels of critical thinking. Moreover, online discussions have been proven to enhance students' elaboration, sharing, and application of knowledge, all important factors of collaborative learning (12).

For discussion boards to be effective in a learning environment, they need to have a specific and dynamic structure. It is critical to have various forums for different topics and instructing students to post several times in response to prompts and to each other's posts focus the discussion board on a student-to-student interaction (13). This means that students can use technology and online tools to forge their learning path, rather than being directly instructed into a specific one. The concept of constructive learning allows for the identification of early misconceptions and biases that were created in initial learning processes that can be reflected upon and modified. This is especially significant in more intermediate to advanced stages of learning (14).

Calderon and Sood propose a content analysis of the posts in an asynchronous online discussion board of three cohorts of graduate students to evaluate and identify the organically emerging indicators of learning and their patterns. For this study, each cohort was divided into groups of three to four students. They would have access to their group's asynchronous discussion board. The instructor would have access to all discussion boards and would act as a discussion facilitator. A case study would be provided weekly, and each group would have to read it and respond to a prompt regarding the case study in their respective discussion boards. The case study entailed topics to be covered in class the following week, and in order to answer the prompts, students would have to go over the weekly assigned readings. This provided a self-monitoring factor, in concomitance to student interaction to the task, aiming to promote the most effective online learning outcomes. The instructor would then read each post and provide feedback to each group, emphasizing their role as facilitators, which, alongside student collaboration, adopted the Community of Inquiry model into the learning setting (10).

The study found that students, on average, responded between 1.7 and 2.5 times per assignment. Moreover, qualitative analysis of each student's post revealed three main themes that represented different learning dimensions and ten learning categories that all fit into the three themes. The first content dimension was contextual communication; students summarized the case study's main points, exposed their diagnostic impressions and the justification behind them, and proposed treatment plans for the case study's patient. Within the realm of contextual communication, the content exposed by the students conveyed a sophisticated level of contextual learning, which was naturally depicted in the discussion posts (10).

The second dimension pertained to inter-personal communication within discussion threads. Students would organically give positive feedback and constructive criticism. This feedback indicated that the students were capable of engaging in complementary critical thinking and communicating respectfully and efficiently. Some students also conveyed leadership in learning within this dimension by encouraging their peers to engage in the discussion (10).

The last dimension observed was unexpected and involved students' self-reflection on how they viewed and understood the content provided to them. The meta-learning dimension presented that students organically contemplated and assessed their own learning in the context of the topic at hand. Students reflected on their learning process and associated the topic of the assignment to their own personal experiences, applying the course's content to settings outside of the class (10).

In conclusion, the use of asynchronous online discussion boards based on a Community of inquiry approach demonstrated that students acquired essential interpersonal communication skills. Peer collaboration naturally took place as learning developed in the online setting. Moreover, the organic self-reflection and meta-learning patterns found within the students' posts bolster the idea that asynchronous online discussion boards, via collaborative learning, promote constructive learning, providing students with the ability to identify learning biases and modify them more effective learning experiences.

Creating a Community of Inquiry through online forum discussions contributes to decreasing the detrimental effects of the big-fish-little-pond effect (BFLPE). This is the concept of decreased academic achievement and self-concept due to comparison to peer achievement. This comparison increases competition and generates a negative impact on learning and academic self-awareness (15). The online forum environment requires cooperation and replaces the competitive learning experience with collaboration, bolstering learning interaction and development. Thus, differences in knowledge are beneficial in this community as it enhances teamwork in learning advancement.

3rd Study: Enhancing online communication to improve learning through self-monitoring and self-regulation

As the academic setting grows and becomes more competitive through time, traditional learning resources become limited and do not effectively fulfill student's needs for positive learning outcomes (16). Based on this context, blended learning strategies, including online components to in-person classes and peer interaction, have been incorporated to achieve beneficial learning results in different subjects effectively. Online and collaborative learning strategies have especially been incorporated into foreign language classes to bolster students' proficiency (17). Withstanding, a study from the University of Wollongong proposed to evaluate if students perceive online collaborative learning as beneficial to their proficiency in writing in French (16).

It is argued that to become proficient in a language, an individual must acquire an extensive collection of verbal and writing skills (18). These skills allow students to effectively engage in social and academic interactions in the target language (19). The concept of social constructivism and, inherently, collaborative learning is widely applied to create a setting for learning about and through language and learning the language itself. Through this concept, students can achieve learning outcomes that they have not been able to do by themselves and thus address their learning shortcomings (20,21). This assistance is known as "scaffolding"; students' interactions with others more experienced than them allow them to understand course content better.

Students' language improvement has widely benefitted from collaborative learning and working in groups. Nonetheless, sometimes an in-person setting can be intimidating and prevent student participation in language development (22). Therefore, online discussions provide greater student involvement and improved argumentation and writing skills in foreign languages (23). It is important to note that learning a foreign language is a dynamic process and obtaining writing proficiency in a foreign language requires "multiple literacies," meaning that students are not only required to know words and grammar in a target language but also have the need to develop critical text analysis skills (24).

In the study by Bissonauth-Bedford and Stace, a blended learning writing project was proposed for students in their third year of French at the University of Wollongong through a collaborative and socio-constructivist approach. Students would be divided into groups and have to write a continuation to a French short story (*L'enfant de sous le pont* by J.M.G. Le Clézio) in the form of a newspaper article. To fulfill this task, students would have to post on online forums and diaries, allowing them to reflect upon their work, introduce a meta-learning component to the task, and comment on their peers' ideas. The students would have five weeks to complete the writing task, in which they would also be answering comprehension and analysis questions about the text in groups on their weekly lectures. Having the writing task be in a newspaper article format encouraged connections with subjects outside of the classroom as most students were journalism majors. In the forum, students were supposed to expose their work progress and reflect upon their progress in writing proficiency. Students were also asked to turn in, along with the group article, an individual reflective summary of their learning and writing improvement and their opinions on group work efficiency for this objective. At the end of the semester, students answered a survey in which they were asked to evaluate their progress and confidence in writing in French and which tasks assisted them in improving their writing skills in French (16).

Study results conveyed that a majority of students believed their writing skills in French had average or good improvement. Moreover, no student believed that their writing skills did not improve at all. Regarding the efficiency of group work in improving writing skills in French, 71% of students thought group work was beneficial for their learning in French. However, 29% of students did not find the group writing task helpful, which emphasizes previous research results that depict students' mixed feelings toward group writing tasks (16).

Students' reflection upon the group writing task contemplates their meta-learning process to writing in a group. Students found that the most useful parts of the group writing task were knowledge-sharing and creating connections between learning a foreign language and other subjects, like journalism. Moreover, students' reflective summaries also exposed their feelings toward peer collaboration in an online and in-person setting. They conveyed that the online setting allowed them to self-monitor their work progress and self-regulate the jobs needed to perform the task. Nonetheless, students also conveyed that communication between peers was better in a face-to-face setting, portraying students' mixed feelings when perceiving a group task (16).

Overall, students perceived group writing tasks as a beneficial tool in improving their writing skills in French. The structure of the writing task and the blended learning model used to carry it out contributed to the students' application of their knowledge in and outside of the classroom. Connecting concepts between subjects improves meta-learning and convey ways in which working in groups may contribute to self-reflection, self-monitoring, and self-regulation of learning strategies and thus reduce implicit biases in learning a foreign language.

The socio-constructivist approach to blended collaborative learning styles seen in this study might suggest a potential to address the lack of instant gratification for in-depth learning, as discussed in Chapter 3. Student communication and collaboration in different mediums increase intrinsic motivation to learning, tying into the concept of delay of gratification, an important feature of self-regulated learning (25). In this case, scaffolding and teamwork function as motivational factors for an ultimate long-term goal in learning, in this case, improved French writing. Interacting with different knowledge levels and cooperating with different groups incentivizes self-reflection and maintaining learning practices.

4th Study: Testing meta-learning techniques in biomedical sciences

When it comes to higher education, a students' previous academic history does not necessarily reflect their ability to excel in university. Even with similar entrance credentials, some students excel while others struggle to meet academic standards. This discrepancy deeply relates to a student's ability to recognize and coordinating their learning (26,27). Not knowing how to regulate one's learning can propagate previously learned cognitive biases into future learning, which might explain how students with similar academic profiles may perform differently in the higher education setting (28). Although higher education is supposed to build the foundation for lifelong learning, many students lack the tools to do so (29). The introduction of meta-learning into higher education has improved academic performance and retained lifelong learning skills (30). This is especially important in the health sciences context, given that knowledge in this area is constantly and rapidly generated so that students can keep up with new information after they graduate (31).

In their study "Prompting undergraduate students' metacognition of learning: implementing 'meta-learning assessment tasks in the biomedical sciences" Colthorpe and colleagues seek to answer the question of whether the implementation of meta-learning tasks that contribute to the self-reflection and regulation of learning help develop immediate and future learning skills in the context of the biomedical sciences (31).

Meta-learning is a term used to describe an individual's understanding of actions and approaches utilized in their learning process (32,1). It is the idea of how one learns and what strategies they are implementing to acquire knowledge. It is a dynamic process that requires self-reflection, self-regulation, and self-monitoring, which allow a person to change their learning strategies based on their current needs and circumstances (1,33). As meta-learning encourages students to be autonomously self-reflective, it allows them to identify cognitive biases that hinder their learning experiences, prompting them to find solutions to deconstruct these biases (28).

An intrinsic part of meta-learning is the process of self-regulation. Self-regulation allows learners to use their metacognitive, behavioral, and motivational strategies to mold their learning (34). Several models for the self-regulation of learning have been implemented and evaluated throughout history, the most common of which is the three-phase model proposed by Zimmerman (35). The three-phase model dictates that effective self-regulation occurs in three stages: 1) the forethought, in which individuals use their self-motivation beliefs to set up strategies and goals to achieve the desired learning outcome; 2) the performance, which is the phase that a student applies their preplanned strategies to learning. This is the phase in which learning effectively occurs; and 3) the self-reflection, in which students utilize self-judgment to assess the learning strategies they employ and justify reasons for errors and success in their learning process (36).

Zimmerman also implies that within the self-regulation model, there are reactive and proactive students. Reactive learners are the ones that generally do not take part in the forethought or self-reflection phases; they focus solely on their performance in acquiring knowledge. On the other hand, proactive learners go through the complete self-regulation cycle and thus better retain acquired knowledge and learning skills for future learning endeavors. Reactive learners fixate their cognitive biases into their immediate learning strategies and thus continue to make mistakes that will impair their acquisition of lifelong learning skills. On the other hand, proactive learners can identify their cognitive biases and make changes further to improve their future learning strategies (36).

Colthorpe et al. propose the introduction of meta-learning tasks within the curriculum of a physiology course in a cohort of 100-400 students for two semesters. Thus, they designed a series of 4-5 meta-learning tasks with 2–3-week intervals between each one that was employed at specific times during the semester to encourage students to become more proactive learners. Each task contained six open-ended questions, often with more than one part. Tasks that were applied early in the semester stimulated students to reflect on recent course content and to set goals and learning strategies for the rest of the semester. Tasks implemented before the mid-semester assessment asked students to use forethought planning strategies to study for the exam. Tasks implemented after the mid-semester exam prompted students to contemplate the learning strategies they used to prepare for the exam and judge whether they were effective or not. The final task of the semester induced students to reflect upon the meta-learning tasks and elaborate on their effect on their learning. Students were also requested to consider adaptations they may have done to their study strategies and justify the changes (31).

When compared to test results of cohorts prior to applying meta-learning tasks in the physiology curriculum, assessment scores of students that answered the meta-learning tasks were significantly higher. However, this result should be interpreted with caution as the meta-learning tasks may not have been the only factor influencing score improvement. Moreover, when asked if the meta-learning tasks were useful to promote learning, 85% of students thought they were useful. Students said the tasks motivated them to be more goal-oriented, aided in time management, stimulated them to review the course's content, and prompted self-reflection and evaluation. On the other hand, the students who believed the meta-learning tasks were not useful reported they already had well-established learning strategies and self-reflection of said strategies and thus considered them unnecessary (31).

When inquired about adaptations made on their learning strategies throughout the semesters, 49% of students reported not making any changes to their learning strategies as they deemed them effective in achieving their desired academic standards.

The other 51% of students reported making adaptations to their learning strategies throughout the semesters, motivated by self-reflection processes generated by the meta-learning tasks. While the assessment scores of the group of students that did not adapt their learning strategies were higher in both semesters than the students that did adapt their methods, students that did adapt their methods had a within-group improvement in assessment scores of 5% in the second semester (31).

The constant turnover in scientific knowledge production in the biomedical sciences elucidates the need for students in higher education pursuing careers in this field to develop effective lifelong learning skills so that they are able to keep up with the overflow of new information and do not become obsolete in the field. Meta-learning has been proven to aid students in creating effective learning strategies for immediate and future learning. Thus, the introduction of self-regulated meta-learning through meta-learning tests can bolster students' learning outcomes and help them create lifelong learning skills.

Given the high influx of knowledge production in the health sciences field, individuals are constantly faced with conflicting knowledge about the same concept. This study provides a possible approach to the issue of cognitive dissonance discussed in chapter 4. The self-regulation techniques employed by students when faced with different learning retention results translate into a more proactive stance toward learning strategies. This allows them to adjust when presented with competitive forces that might interfere with their learning progress, even making it possible for them to incorporate these forces into their new learning strategies. In fact, when students apply the self-reflection phase of Zimmerman's model to their learning process, they can utilize cognitive dissonance to mold further their new learning strategies based on the new information being presented to them.

5th Study: Using flipped learning to improve learning (measured as concept threshold)

Every subject has a more complex concept that is harder for students to grasp while learning its content. These are known as threshold concepts, and they require adaptive forms of learning for students to understand them (37,38) effectively. Meyer and Land created a set of aspects needed to overcome threshold concepts, including the need for students to be able to explain the concept and apply it in different areas of their life to convey a complete understanding of the concept, the need for peer assessment so students can share their doubts and anxieties and understand that they are not alone in the learning process, and they need a recursive and excursive approach to their learning, meaning that they need to take in different views on the concept and that their learning process allows topic divergences, respectively (38). Students then require self-efficacy measures to overcome threshold concepts as they become aware of effective learning strategies and put them in the appropriate frame of mind to conquer said concepts (39). Active learning is thought of being the most effective form of overcoming threshold concepts. However, which methods of active learning are best to tackle these concepts is still unclear. Therefore, Olaniyi proposes a study to answer the question *of which methods can help students overcome threshold concepts*. The author introduces flipped learning as a possibility of one of these methods (40).

In flipped learning, the classroom and at-home work are inverted, meaning students learn the course's content at home and apply it to practice in the classroom (41). This teaching model allows for greater peer interaction during class time and allows teachers to work as facilitators and focus on the main areas and students who need it more (42). Flipped learning utilizes a myriad of tools, online and in-person, to help students improve their critical thinking and problem-solving skills and promote active learning (43). These tools can include interactive videos, online tests and quizzes, and peer interaction (43). This student-centered learning method prompts students to self-monitor and self-regulate their studying, given that they are introduced to a concept’s content individually on their own time (45). Meta-learning is thus encouraged in the flipped learning method as it will promote the identification of effective learning strategies for each student and allow them to take responsibility for their learning (46,47,48).

Olaniyi's study, "Threshold concepts: designing a format for the flipped classroom as an active learning technique for crossing the threshold" proposes a flipped learning model to overcome the threshold concept of thermodynamics in physics. This study was carried out with three groups taking physics classes over two academic years. The flipped learning method occurred during four weeks, with three-hour lessons occurring once a week. Students were provided with reading materials, interactive videos, guiding questions, and other tools on the subject of thermodynamics before each lesson. They were asked to study the material to participate in the activities that would take place during the lessons. The teacher served as a facilitator, answering doubts or concepts that were left unclear. Students were also surveyed about their perspective on the flipped learning method and if it helped them better understand the concept of thermodynamics. The effectiveness of the flipped learning style would also be assessed through the student's performance on the semester exam for thermodynamics (40).

The majority of students agreed that flipped learning helped them to understand the concept of thermodynamics better. Moreover, they conveyed positive feedback on them taking charge of their own learning, depicting that meta-learning and self-regulation strategies took place in the context of the flipped learning model. As a result, students became more independent and learned what works best for them when learning a threshold concept. In one of the cohorts, 19 students (out of 26) passed the exam, and of those 19, only 5 students did not try to answer the questions on thermodynamics, meaning more than 70% of students did well on this topic on the exam. In the second group of students, all 45 students taking the class passed the exam, and only 5 students did not attempt to answer the thermodynamics questions. And in the third cohort, out of 27 students, only 2 did not attempt to do the thermodynamics questions, with 25 students doing well on this topic (40).

The results of Olaniyi's study convey that flipped learning is an effective strategy to help students overcome threshold concepts. This model encourages meta-learning and self-regulated learning as students take responsibility for their studying and become aware of the best methods to approach the flipped learning activities. Meta-learning is also essential to help students understand threshold concepts, especially when combined with peer interactions and excursive and recursive approaches to learning. Therefore, using a flipped learning model to tackle threshold concepts provides students with the tools they need to understand the concept and move forward with their learning.

Conclusions

Meta-learning is an intrinsic tool needed to enhance an individual's learning outcomes while helping them acquire lifelong learning skills. Whether it be directly or indirectly, meta-learning can be organically stimulated in different learning methods, including blended and flipped learning, as well as in collaborative learning. It can be generated through concepts such as constructivist learning and the creation of a community of inquiry. Its introduction should be encouraged at the earlier stages of cognitive learning, as meta-learning can help students identify their cognitive biases and prompt them to adapt their learning strategies to reduce or eliminate them.

References for appendix 1 chapter

1. Sadler-Smith, E., Evans, C., Boström, L., & Lassen, L. M. (2006). Unraveling learning, learning styles, learning strategies and meta-cognition. *Education+ Training*.

2. Boström, L. (2012). Do Ten-Year-Old Children in Sweden Know How They Learn? A Study of How Young Students Believe They Learn Compared to Their Learning Styles Preferences. *International Education Studies*, *5*(6), 11-23.
3. Dunn, R. Dunn, K., & Price, G. E. (1984, 1991, 2000, 2001). Productivity Environmental Preference Survey. Lawrence, KS: Price System.
4. Schinske J, Tanner K. Teaching more by grading less (or differently). CBE—Life Sciences Education. 2014 Jun;13(2):159-66.
5. Allen, I. E., & Seaman, J. (2013). *Changing course: Ten years of tracking online education in the United States*. Sloan Consortium. PO Box 1238, Newburyport, MA 01950.
6. Allen, I. E., & Seaman, J. (2016). *Online report card: Tracking online education in the United States*. Babson Survey Research Group. Babson College, 231 Forest Street, Babson Park, MA 02457.
7. Calderon, O. (2013). Preferred Features of Course Management Systems in Post Secondary and Corporate Online Learning. In *Online Instruments, Data Collection, and Electronic Measurements: Organizational Advancements* (pp. 311-327). IGI Global.
8. Rovai, A. P. (2002). development of an instrument to measure classroom

community. *The Internet and higher education*, *5*(3), 197-211.
9. Rovai, A. P., Wighting, M. J., & Lucking, R. (2004). The classroom and school community inventory: Development, refinement, and validation of a self-report measure for educational research. *The Internet and Higher Education*, *7*(4), 263-280.
10. Calderon, O., & Sood, C. (2020). Evaluating learning outcomes of an asynchronous online discussion assignment: a post-priori content analysis. *Interactive Learning Environments*, *28*(1), 3-17.
11. Ke, F. (2013). Online interaction arrangements on quality of online interactions performed by diverse learners across disciplines. *The Internet and Higher Education*, *16*, 14-22.
12. Ioannou, A., Demetriou, S., & Mama, M. (2014). Exploring factors influencing collaborative knowledge construction in online discussions: Student facilitation and quality of initial postings. *American Journal of Distance Education*, *28*(3), 183-195.
13. Brooks, C. F., & Bippus, A. M. (2012). Underscoring the Social Nature of Classrooms by Examining the Amount of Virtual Talk across Online and Blended College Courses. *European Journal of Open, Distance and E-learning*.

14. Ertmer, P. A., & Newby, T. J. (1993). Behaviorism, cognitivism, constructivism: Comparing critical features from an instructional design perspective. *Performance improvement quarterly*, *6*(4), 50-72
15. Fang J, Huang X, Zhang M, Huang F, Li Z, Yuan Q. The big-fish-little-pond effect on academic self-concept: a meta-analysis. Frontiers in psychology. 2018 Aug 29;9:1569.
16. Bissoonauth-Bedford, A., & Stace, R. (2015). Building a writing community through learning of French. *Journal of University Teaching & Learning Practice*, *12*(2), 7.
17. Bissoonauth-Bedford, A., & Stace, R. (2012). Grappling with grammar on a Virtual Learning Platform: The Case of First-Year French Students at the University of Wollongong. *Journal of University Teaching and Learning Practice*, *9*(1), 2.
18. Peregoy, S. F., & Boyle, O. S. (1997). Reading, writing and learning in ESL: A resource book for K-12. *White Plains, NY: Longman*.
19. Van Nguyen, L. (2010). Computer-mediated collaborative learning within a communicative language teaching approach: A sociocultural perspective. *The Asian EFL*

Journal Quarterly March 2010 Volume 12, Issue, 202.
20. Laurillard, D. (2009). The pedagogical challenges to collaborative technologies. *International Journal of Computer-Supported Collaborative Learning*, *4*(1), 5-20.
21. Vygotsky, L. S. (1980). *Mind in society: The development of higher psychological processes*. Harvard university press.
22. Kroonenberg, N 1995. The French connection: Public journals for high school students. In Warschauer, M (ed.), Virtual Connections: Online activities and projects for networking language learners, Second Language Teaching and Curriculum Centre, University of Hawai'i, Honolulu, pp. 27-29.
23. Kern, R. G. (1995). Restructuring classroom interaction with networked computers: Effects on quantity and characteristics of language production. *The Modern language journal*, *79*(4), 457-476.
24. Allen, H. W. (2009). A multiple literacies approach to the advanced French writing course. *The French Review*, 368-385.
25. Bembenutty H, Karabenick SA. Inherent association between academic delay of gratification, future time perspective, and self-regulated learning. Educational psychology review. 2004 Mar;16(1):35-57.

26. Dunlosky, J., Rawson, K. A., Marsh, E. J., Nathan, M. J., & Willingham, D. T. (2013). Improving students' learning with effective learning techniques: Promising directions from cognitive and educational psychology. *Psychological Science in the Public Interest*, *14*(1), 4-58.
27. Mega, C., Ronconi, L., & De Beni, R. (2014). What makes a good student? How emotions, self-regulated learning, and motivation contribute to academic achievement. *Journal of educational psychology*, *106*(1), 121.
28. Griffiths, T. L., Callaway, F., Chang, M. B., Grant, E., Krueger, P. M., & Lieder, F. (2019). Doing more with less: meta-reasoning and meta-learning in humans and machines. *Current Opinion in Behavioral Sciences*, *29*, 24-30.
29. Boud, D., & Falchikov, N. (2006). Aligning assessment with long-term learning. *Assessment & evaluation in higher education*, *31*(4), 399-413.
30. Biggs, J. (1988). The role of metacognition in enhancing learning. *Australian Journal of education*, *32*(2), 127-138.
31. Colthorpe, K., Sharifirad, T., Ainscough, L., Anderson, S., & Zimbardi, K. (2018). Prompting undergraduate students' metacognition of learning: implementing

'meta-learning assessment tasks in the biomedical sciences. *Assessment & Evaluation in Higher Education*, *43*(2), 272-285.
32. Jackson*, N. (2004). Developing the concept of metalearning. *Innovations in Education and Teaching International*, *41*(4), 391-403.
33. Winters, T. (2011). Facilitating meta-learning in art and design education. *International Journal of Art & Design Education*, *30*(1), 90-101.
34. Puustinen, M., & Pulkkinen, L. (2001). Models of self-regulated learning: A review. *Scandinavian Journal of Educational Research*, *45*(3), 269-286.
35. Zimmerman, B. J. (2000). Attaining self-regulation: A social cognitive perspective. In *Handbook of self-regulation* (pp. 13-39). Academic Press.
36. Zimmerman, B. (2002). Becoming learner: Self-regulated overview. *Theory into Practice*, *41*(2), 64-70.
37. Cousin, G. (2006). An introduction to threshold concepts.
38. Meyer, J., & Land, R. (2003). *Threshold concepts and troublesome knowledge: Linkages to ways of thinking and practising within the disciplines* (pp. 412-424). Edinburgh: University of Edinburgh.

39. Rhem, J. (2013). Before and after students “get it”: threshold concepts. In *National Teaching and Learning Forum Newsletter* (Vol. 22, No. 4).
40. Olaniyi, N. E. (2020). Threshold concepts: designing a format for the flipped classroom as an active learning technique for crossing the threshold. *Research and Practice in Technology Enhanced Learning*, *15*(1), 2.
41. Khoo, E., Peter, M., Scott, J., & Round, H. (2018, June). Flipped classroom learning in a first-year undergraduate engineering course. In *EdMedia+ Innovate Learning* (pp. 1275-1280). Association for the Advancement of Computing in Education (AACE).
42. University of Plymouth. (2014). 7 steps to a flipped classroom. University of Plymouth: Educational Development Teaching and Learning. Online resource. https://www.plymouth.ac.uk/uploads/production/document/path/2/2399/7_Steps_to_a_Flipped_Classroom.pdf
43. Hwang, G. J., Lai, C. L., & Wang, S. Y. (2015). Seamless flipped learning: a mobile technology-enhanced flipped classroom with effective learning strategies. *Journal of computers in education*, *2*(4), 449-473.
44. Bialik, M., & Fadel, C. (2015). Meta-learning for the 21st century: What should

students learn. *Center for Curriculum Redesign. https://curriculumredesign.org/wp-content/uploads/CCR-Meta-Learning-FINAL-Nov.-17-2015. pdf.*
45. Sun, X. L. (2017, September). An action research study from implementing flipped classroom model in professional English teaching and learning. In *3rd Annual International Conference on Social Science and Contemporary Humanity Development (SSCHD 2017)*. Atlantis Press.
46. Galster, M., Mitrovic, A., & Gordon, M. (2018). Toward enhancing the training of software engineering students and professionals using active video watching. In 2018 IEEE/ACM 40th International Conference on Software Engineering: Software Engineering Educa
47. JISC, E., & Team, L. (2010). Effective assessment in a digital age-a guide to technology-enhanced assessment and feedback. *Technology-enhanced Assessment*, 26-28.
48. Langbauer, M., & Lehner, F. (2015, October). An interactive video system for learning and knowledge management. In *2015 International Conference on Enterprise Systems (ES)* (pp. 55-65). IEEE.

Epilogue: The journey to deep and effective learning

The brain was designed to learn, save resources, and act efficiently. If brains were given a choice, we may wonder which would be preferable, a degree from an Ivy League university with little learning, such as cramming the day before the exam and finding strategies to pass with minimal effort, or acquiring the same diploma through hard work. It is likely that our survival mode would come into action and select the method that requires less energy.

This does not make our brain wrong. The brain is simply doing what thousands of years of evolution did for survival: saving energy to increase survival. However, this evolution does not lead to learning less deeply. Rather, this is caused by people and traditional education. This can be seen through systems such as the Ivy League universities mentioned above. The benchmarks for learning and professional success are based on imperfect systems. The paradox is that people created a further problem in order to organize and label education: assigning incorrect benchmarks for deep learning. With measurements of learning, the school system created the false impression that the topic has been learned sufficiently for practicing a profession once a diploma or degree has been acquired. However, diplomas and degrees could be obtained through superficial learning.

Interestingly, society admits that this system is flawed; otherwise, there would be no need for board exams beyond the university system, such as in medicine and law. In addition, there are often examples of below-average performance in various professions.

This, in a way, parallels medicine. Medicine usually addresses immediate and local problems, while rarely addressing the main issue of the body losing its capacity for balance and healing. For instance, doctors prescribe anti-inflammatory medication for patients with severe inflammation. However, they usually do not go deep into the problem of body imbalance, which leads to excessive inflammation.

A similar situation can be seen in the field of education. Teachers prescribe books, extra hours of tutoring, and memorization to students with poor grades to help improve their grades in standardized tests. However, teachers do not usually address the fundamental problem regarding the reasons behind students failing to learn deeply.

This book did not seek to address all potential causes of the lack of deep learning but rather examine our hidden cognitive biases, which likely exacerbate this problem. This book defined this as the paradox of learning, as we may think we are learning when, in fact, we are not. The first crucial step in prevention is recognition, as discussed in the final chapters. However, this is a multifaceted problem, including issues of the traditional educational system. It is my hope that this book will inspire readers to consider and reexamine this topic and share their thoughts and opinions. I would be delighted to hear others' thoughts and exchange ideas (my email address is given below).

Sincerely,

Felipe Fregni (felipe.fregni@medicaltli.org)

Boston, Massachusetts

About the Author

FELIPE FREGNI, M.D., Ph.D., M.M.Sc., M.P.H, M.Ed., is a Professor of Physical Medicine & Rehabilitation at Harvard Medical School and a Professor of Epidemiology at Harvard T.H. Chan School of Public Health. He has designed and has been leading one of the most successful interactive online programs in clinical research: the Principles and Practice of Clinical Research offered currently by Harvard T.H. Chan School of Public Health. He leads a large research center at Spaulding Rehabilitation Hospital that investigates the neural mechanisms of learning in rehabilitation.

Made in the USA
Monee, IL
19 April 2026

48628914R00104